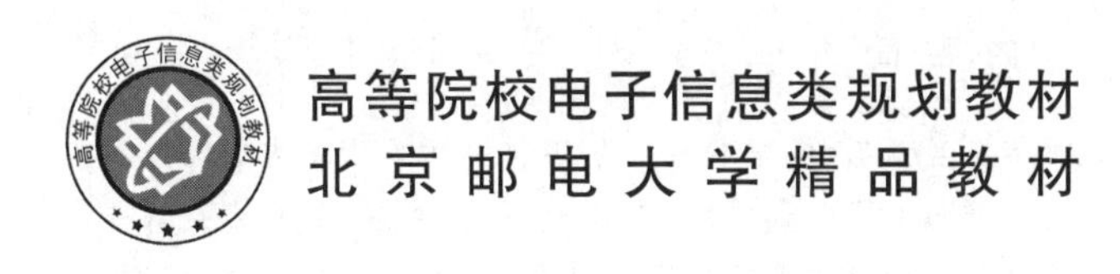

通　信　英　语

（第 7 版）

ENGLISH ON TELECOMMUNICATIONS

石方文　葛子刚　编著

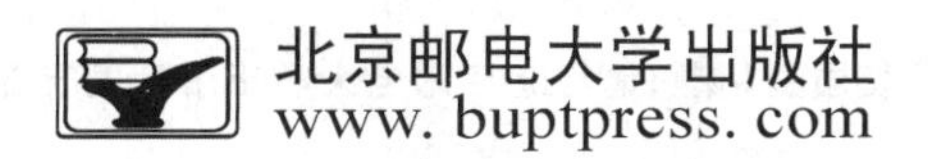

内容简介

《通信英语》(第7版)是通信信息领域的专业英语教材,全书共分为17个单元。书中内容涵盖了当下重要的信息新技术领域,包括移动通信、光传送网OTN、云技术、物联网、人工智能AI、宽带接入技术和区块链等。本书的课文主要选自国外高等学校的教科书和一些高级别的相关技术刊物。这些课文语言朴实,文字流畅,易于阅读和理解。由于这些文章的作者背景不同,写作风格各异,用词和语法运用也各有侧重,因而对于读者专业英语水平的提高定会有很大的帮助。

图书在版编目(CIP)数据

通信英语 / 石方文,葛子刚编著. -- 7版. -- 北京:北京邮电大学出版社,2021.4(2023.8重印)
ISBN 978-7-5635-6364-7

Ⅰ.①通… Ⅱ.①石… ②葛… Ⅲ.①通信—英语—教材 Ⅳ.①TN91

中国版本图书馆CIP数据核字(2021)第069684号

策划编辑:彭 楠 **责任编辑**:王晓丹 左佳灵 **封面设计**:七星博纳

出版发行:北京邮电大学出版社
社 址:北京市海淀区西土城路10号
邮政编码:100876
发 行 部:电话:010-62282185 传真:010-62283578
E-mail:publish@bupt.edu.cn
经 销:各地新华书店
印 刷:唐山玺诚印务有限公司
开 本:787 mm×1 092 mm 1/16
印 张:15.5
字 数:364千字
版 次:1994年6月第1版 1996年6月第2版 2001年10月第3版 2004年7月第4版
2008年1月第5版 2014年8月第6版 2021年4月第7版
印 次:2023年8月第3次印刷

ISBN 978-7-5635-6364-7 定价:39.00元

编者的话

为了适应我国通信信息产业飞速发展的新形势，提高工程技术人员的专业英语水平，我们编写了这本《通信英语》。

本书共17个单元，内容主要涉及数字通信、移动通信、云计算、物联网、宽带接入技术、人工智能、区块链、IPv6、软件定义网络等，基本覆盖了当代通信的重要技术领域。

本书的课文主要选自美国高等学校的教科书和一些高级别的英文通信技术刊物。这些课文语言朴实，文字流畅，易于阅读和理解。同时，由于文章出自不同的作者，体现了不同的文风和文体，极有助于拓展和提高读者的视野和阅读能力。因此，相信本书会受到读者的欢迎。

考虑到读者学习过公共英语，已具有一定的英语基础，所以本书的编写是以扩大读者通信技术方面的英语词汇量，使读者熟悉专业术语、了解科技文章的表达特点和掌握英语翻译技巧为宗旨的。我们认为，只要读者能熟练地阅读和翻译本书的课文，看一般的通信专业英语文章就不会感到费力。

衷心希望本书的出版能为我国IT业的发展，为通信信息产业职工整体素质的提高，为我国高等教育的发展做出贡献。

石方文
2020年于北京邮电大学

第 7 版前言

《通信英语》已经走过了 27 年的历程，它的第 7 版也要很快问世。

书，尤其是教科书的生命力，首先在于读者的认同，在于学生与教师的认同。自 1993 年始，《通信英语》先后作为邮电系统工程技术人员继续教育的指定用书、计算机通信专业（本科）和通信信息管理专业（专科）的全国自学考试用书、全国邮电函授教育和北京邮电大学网络教育的教材、国内多所大学本科和研究生的教学用书，读者群体相当庞大，印刷册数不断增加。作者以为，这表明《通信英语》已得到人们的广泛认可。

专业英语类教科书的生命力，又在于它的语言。由于《通信英语》的课文主要选自美国高校的教科书和一些全球性的高级别英语技术刊物，因而句法地道，语言朴实，文字流畅。同时，由于课文出自不同作者之手，他们的用词和语法习惯又不尽相同，因而展示了不同的文风和文体，能够较全面地拓展和提高读者的视野和英文水平。作者以为，这正是《通信英语》深受好评的原因所在。

专业英语类教科书的生命力，还在于它的专业性。《通信英语》的课文，几乎涵盖了当代通信信息行业的所有新技术领域。在每个版本问世之前，我们都要重新筛选课文，删除技术已经落伍的文章，添加新技术的内容，使教材始终跟随信息领域的最新潮流。在这次修订时，我们又对原书做了较大的调整，增加了诸如光传送网 OTN、5G、6G、人工智能、边缘计算、物联网、IPv6 等专题，从而使教材更加适应电信和信息技术的今天和未来。我们精心选择每篇文章，注重专业的基本概念，强调原理描述的深入浅出，从而使读者更易集中精力，专注英语学习，这亦是本书长盛不衰的重要原因。

对 20 多年来支持本书出版、积极推广和使用本书的单位和朋友们表示衷心的谢意！

石方文

2020 年 7 月

目　　录

第一部分

教学大纲

《通信英语》教学大纲

一、课程性质与设置目的

“通信英语”是高等学校通信信息类专业的一门专业基础课。

随着我国通信信息产业的迅猛发展，提高工程技术人员的专业英语水平成为当务之急。设置“通信英语”课程，就是为了扩展学生在通信信息专业方面的英语词汇量，熟悉该领域的专业术语，了解科技英语的表达特点，掌握专业英语的翻译技巧，从而大大提高他们的业务素质并增强他们在专业上与国际接轨的能力。

二、课程的基本内容

本课程选用教材为《通信英语》(第 7 版)，由石方文、葛子刚编著。

《通信英语》(第 7 版)共 17 个单元，内容涉及光传送网 OTN、云计算、物联网等方面，基本覆盖了当代通信的所有新技术领域。

《通信英语》(第 7 版)的课文主要选自美国的一些高级别的通信技术刊物。这些课文语言朴实，文字流畅，易于阅读和理解。由于文章作者不同，写作风格各异，语法现象也各有侧重，因而对于读者专业英语水平的提高，定会有所帮助。

三、课程的基本要求

1. 词汇要求

(1) 对课文中出现的词汇能说出其汉语词义。

(2) 对课文中常见的通信专业词组要能够英汉互译，并拼写正确。

2. 语法要求

应熟练掌握下述常见的语法现象：

(1) 动词不定式；

(2) 动名词；

(3) 现在分词；

(4) 过去分词；

(5) 被动语态；

(6) 常用介词；

(7) 各类从句。

3. 阅读要求

能读懂课文或与课文难易程度相当的课外(通信信息专业)英语材料，生词不超过所读材料词数的 2%，速度达每分钟 30 个词，对课文的理解基本正确。

4. 翻译要求

能将阅读的材料译成汉语,译文文字通顺,意思基本准确,笔译速度达到每小时500个英语单词。

5. 听、说和写作要求

可根据学生层次提出不同要求。

四、学习方法

1. 要认真掌握一些常见的语法现象

一些语法现象在专业英语文章中出现的频度很高,例如:The signal received from a satellite, located in far outer space, is very weak.

在这个很短的句子中,过去分词就出现过两次。因此,建议学生在第一次遇到这类语法现象时,就将其作用和翻译方法真正掌握,这样才能将文章理解好,翻译好。

2. 要掌握一些常见的词干、词头和词尾

词干是一个词的躯体,是不变化的部分,体现了该词的基本含义;词头是词干之前的部分,它能赋予该词干以新的词义,派生新词;词尾是词干之后的部分,它也能赋予该词干以新的词义,派生新词。所以,只要掌握了常用的词干、词头和词尾,就能大大地扩展自己的词汇量。

例如,英语中的-duc-或-duct-都是词干,该词干的意义为"引导",如果配以不同的词头和词尾,便衍生出许多英语单词,例如:

conduct *v.* 传导

conductor *n.* 导体

deduce *v.* 推演

deduct *v.* 扣除;演绎

induce *v.* 引起;引诱

inductance *n.* 电感

introduce *v.* 引入;介绍

produce *v.* 生产

reduce *v.* 减少

reproduce *v.* 再生产

再例如,trans-是一个词头,该词头的意义是"横过、跨过、贯通",因而不难判断下述生词的意思:

transcontinental *a.* 横跨大陆的

transfer *v.* 转移;转换;传送

transmit *v.* 发送;传输

transmitter *n.* 发送器;发信机

transport *v.* 运输;运送

translate *v.* 转换;翻译

transplant *v.* 移植;移种

transform *v.* 转换;变换

3. 要精读课文

教材中所选的许多课文都是较典型、地道的专业英语文章。这些文章用词严谨,语法现象规范,专业词组出现频度很高。因此,只要认真地精读几篇文章,即不仅将语法弄懂,还将词汇熟记在心,英语水平就会大有长进。

五、学时分配

本课程学习的总学时为216学时,即18周。学习进度大约每周一课。

六、考试

考试方式为闭卷,笔试,考试时间为2小时。不允许查字典。

评分用百分制,60分为及格。

七、参考书目

《实用英语语法》(第三次修订本),张道真编,商务印书馆,1992。

《新英汉词典》,上海译文出版社,1988。

《网络英语》,北京邮电大学出版社,张筱华、石方文编著。

《信息英语》,北京邮电大学出版社,张筱华编著。

第二部分
主要语法现象

《通信英语》主要语法现象

一、动名词

动名词的作用相当于名词，它在句子中可做主语、宾语或介词宾语。例如：

(1) Reading is a good habit.

动名词在句子中做主语。

(2) I enjoy working with you.

动名词在句子中做宾语。

(3) I am interested in reading.

动名词在句子中做介词宾语。

动名词的作用虽然相当于名词，但它仍然具有一些动词的特征。例如，它可以有自己的宾语和状语。请看下面的例句：

(4) I am interested in carefully reading books.

(5) He insisted on doing it in his own way.

(6) What would you suggest for overcoming the difficulties?

在《通信英语》中，动名词的应用相当普遍，再请看下面的例子：

(7) Digital transmission provides a powerful method for overcoming noisy environments.

(8) Consequently there is an inherent advantage for overcoming noisy environments by choosing digital transmission.

(9) The telephone contains a transmitter and receiver for converting back and forth between analog voice and analog electrical signals.

二、现在分词

1. 现在分词一般用作形容词，可修饰动作的发出者，有主动的意义。可译成"……的……"，此时多做定语。例如：

(1) The smiling girl is Wangling.

(2) This is an amusing story.

在更多情况下，我们可以用现在分词短语做定语，在功能上和一个定语从句差不多。例如：

(3) The girl standing by the window talked to her friend.

(4) There are a lot of boys on the sports ground playing football.

(5) They built a highway leading into the mountains.

在《通信英语》中，现在分词的应用相当普遍，我们来看下面的例子：

(6) Furthermore, we shall prove that a minimum theoretical sampling frequency of order 6.8 kilohertz is required to convey a voice channel occupying the range 300 Hz to 3.4 kHz.

(7) Thus the output to the codec may be seen as a sequence of 8 pulses relating to channel 1, then channel 2, and so on.

(8) Fig. 3-1 illustrates a typical full-duplex data transmission system including the originating data processing equipment and the interface assembly which consists of buffer and control units.

(9) In addition to defining standards covering the NNI, CCITT also embarked on a series of standards governing the operator of synchronous multiplexers and SDH Network Management.

(10) The unquestionable success of the first generation of paging systems using wired networks with signaling devices was followed in the early 1950s by technological progress in the miniaturization of electronic circuitry.

2. 现在分词(分词短语)还可以做状语,用来表示方式、目的、条件、结果和背景,等等;有时表示主语在做一个动作的同时还进行的另一动作,此时称之为伴随状态。例如:

(1) Every evening they sit in sofa watching TV.

现在分词起伴随作用。

(2) I read the English book, using my dictionary.

现在分词表示条件或背景。

还可以列出一些例子:

(3) I ran out of the house shouting.

(4) I got home, feeling very tired.

(5) Not knowing her address, we couldn't get in touch with her.

(6) She got to know them while attending a conference in Beijing.

(7) By comparison, most other forms of transmission systems convey the message information using the shape, or level of the transmitted signal; parameters that are most easily affected by the noise and attenuation introduced by the transmission path.

(8) At the receiving end of an asynchronous serial data link, the receiver continually monitors the line looking for a start bit.

(9) Generally, this link has lower capacity than the total number of PBX subscribers, reflecting the fact that, at any given time, only a fraction of the subscribers will be engaged in external calls.

(10) Today, certain models of pagers are even equipped with a display capability of providing optical messages consisting of 10 to 80 characters, and thus substantially increasing the efficiency of the system.

三、过去分词

1. 过去分词(短语)一般用作形容词,可修饰动作的对象,有被动的意义。可译成"被……的……"或"由……的……",此时多做定语。例如:

(1) The trees planted by me have grown up.

(2) We've already met the target set in the program.

(3) What's the language spoken in this area?

(4) Suddenly there appeared a young woman dressed in green.

(5) For example, the signal received from a satellite, located in far outer space, is very weak and is at a level only slightly above that of the noise.

(6) Once the start bit has been detected, the receiver waits until the end of the start bit and then samples the next *N* bits at their centers, using a clock generated locally by the receiver.

(7) Paging solutions to the above communication situations were introduced in the first half of this century by means of simple optical and acoustical signaling devices installed on the premises of enterprises, hospitals, offices, etc.

(8) The ideal mobile telephone system would operate within a limited assigned frequency band and would serve an almost unlimited number of users in unlimited areas.

(9) This is actually a collection of national networks interconnected to form an international service.

(10) In addition to the five classes of switching centers listed above, the network is augmented with additional switching nodes called tandem switches.

2. 过去分词还可以做状语,例如:

(1) Although originally designed and implemented to service analog telephone subscribers, it handles substantial data traffic via modem, and is gradually being converted to a digital network.

(2) Compared with you, we still have a long way to go.

(3) Moved by their speech, we were momentarily at a loss what to say.

(4) He returned us the papers uncorrected.

四、动词不定式

在《通信英语》中,动词不定式大量地用来做状语或定语。

1. 动词不定式在句首或句末常用来表示目的,可译成"为了……""以便……",可见这时候起目的状语作用。例如:

(1) To do a good job, we use good tools.

(2) We use good tools to do a good job.

(3) We must do everything we can to help them.

由于英语中常用被动语态,所以上面的例(2)可表示为:

(4) Good tools are used to do a good job.

注意此例中的下划线部分,这类句型在《通信英语》中大量出现。类似的句型还有:

…is used to do…

…is needed to do…

…is required to do…

…is arranged to do…

请看下面的例子:

(5) A circuit of the demultiplexer is arranged to detect the synchronization word, and thereby it knows that the next group of 8-digits corresponds to channel 1.

(6) Furthermore, we shall prove that a minimum theoretical sampling frequency of order 6.8 kilohertz is required to convey a voice channel occupying the range 300 Hz to 3.4 kHz.

(7) As each incoming bit is sampled, it is used to construct a new character.

(8) Timing signals from the interface assembly at the transmitter are applied to the data modem to synchronize the computer and the data set.

(9) The switching centers are linked together by trunks. These trunks are designed to carry multiple voice-frequency circuits using either FDM or synchronous TDM.

(10) The 10 regional centers are meshed together to provide full connectivity.

2. 动词不定式在名词之后常用来表示定语,译成"……的……"。例如:

(1) She usually has a lot of meetings to attend in the evenings.

(2) I want to get something to read during the vacation.

(3) She was the first person to think of the idea.

(4) This will be a good opportunity to exchange experience.

(5) The operator of a local telephone switchboard had an option to transmit an optical or acoustical signal for the searched person over a wired network.

(6) Why 800 MHz? The FCC's decision to choose 800 MHz was made because of severe spectrum limitations at lower frequency bands.

(7) Any communication systems which can provide customers with means to communicate from any place, at any time and at any distance gives social and economic advantages.

五、一个重要的语法现象——被动语态

被动语态在《通信英语》中被大量使用,这里给出几个例子:

(1) If we consider binary transmission, the complete information about a particular message will always be obtained by simply detecting the presence or absence of the pulse.

(2) Noise can be introduced into transmission path in many different ways: perhaps via a nearby lightning strike, the sparking of a car ignition system, or the thermal low-level noise within the communication equipment itself.

(3) An asynchronous serial data link is said to be character oriented, as information is transmitted in the form of groups of bits called characters.

(4) When the received character has been assembled, its parity is calculated and compared with the received parity bit following the character.

(5) The first low-loss silica fiber was described in a publication which appeared in October of 1970. The date of this publication is sometimes cited as the beginning of the era of fiber communication.

当然,在《通信英语》中还有许多其他重要的一些语法现象,请读者参阅作者建议的相关书籍,这里不再一一赘述。

第三部分

课　文

UNIT 1

PASSAGE

Optical Transmission Network OTN[1]

In recent years, the services carried by communication networks have undergone tremendous changes.[2] Data services developed rapidly, especially broadband services, such as IPTV and video development. The data service puts forward new requirements to the transmission network. Transmission networks must provide high bandwidth to accommodate this growth. More importantly, it requires a transport network that can be quickly and flexibly scheduled to meet the needs of the service. Currently the main transmission network is using SDH and WDM technology, but they both have some limitations.

SDH technology emphasizes electrical-layer service, and it has flexible scheduling, management and protection mechanism and perfect OAM function. The traditional SDH transmission network has the service scheduling capability of small particle and limited transmission capacity. It cannot transmit efficiently broadband service of large particle. WDM technology based on optical layer has the advantage to provide high capacity of transmission due to its transmission characteristics of multi-wavelength channel.[3] However, the current WDM networks are mainly applied in point to point mode, lacking of effective network maintenance and management tool.[4] Optical scheduling system (such as ROADM) can achieve the scheduling and protection ability similar to SDH, but it is difficult to be applied to a wide range of networks due to the limitation of physical layer and wavelength.

OTN technology has both of the advantages of traditional SDH and WDM. OTN adopts WDM technology in optical layer, which can realize the transmission of large particles.[5] OTN uses asynchronous mapping and multiplexing at the electrical layer to support ODU*k* (k = 1, 2, 3, 4, flex) crossing-connection particles. Compared with SDH VC-12/VC-4 scheduling particles, the transmission particles of OTN multiplexing, crossover and configuration are significantly larger, which can significantly improve the adaptability and transmission efficiency of high-bandwidth data customer services.[6]

The OTN technology includes a complete system structure of optical layer and electrical layer, and each layer has a corresponding mechanism of network management, monitoring and survival. The OTN technology can provide powerful OAM capabilities, and enable up to six tandem connection monitoring (TCM) capabilities, providing advanced performance and fault monitoring.[7] The OTN equipments based on cross-functional ODU*k* can complete circuit switching particles from the SDH-155M to 2.5G/10G/40G, realizing flexible scheduling and protection to large-particles services.[8] The OTN equipments can also introduce ASON-based intelligent control plane, improving the flexibility and survivability of network configuration.[9]

The OTN equipment shall have the functions of customer interface, interface adapter and line interface processing. There are several forms of OTN equipments, such as OTN terminal multiplexer, OTN electrical-crossing equipment, and OTN photoelectric hybrid-crossing equipment.

OTN terminal multiplexer is the WDM equipment that supports the interfaces of OTN, which include the line interface and tributary interface (also called services interface or inter-domain network interface).[10] Performance and fault monitoring of end-to-end wavelength path can be realized. The current mainstream manufacturers of WDM system have basically adopted OTN structure, and support OTN interface in G.709 standard, so they are all OTN terminal multiplexers.

Similar to the existing SDH cross-connection equipment, OTN electric-crossing equipment improves the crossing function of circuit ODU*k* level, providing flexible circuit scheduling and protection for OTN network.[11] OTN electric-cross devices can exist independently, similar to the SDH DXC equipments, providing a variety of external interfaces and OTU*k* interfaces.[12] They can also be integrated with the OTN terminal multiplexer to provide optical multiplexing and optical transmission section function supporting WDM transmission.

Electrical-crossing equipment of OTN can combine with OCh crossing equipment (ROADM or PXC) to provide scheduling function of ODU*k* electrical layer and OCh optical layer. Wave-level services span OCh directly, and other services that need to be scheduled must cross over ODU*k*.[13] The combination of both can complement each other, and avoid their weaknesses. This large-capacity scheduling equipment is photoelectric hybrid cross of OTN equipment.

Typical equipment of this type is OSN6800 produced by Huawei, which supports OCh-crossing based on ROADM and ODU1 / GE crossing. ODU1 crossing capacity is 320 GB and supports ASON control plane. Other manufacturers, such as ZTE, are developing similar products.

According to the different existing forms of OTN equipments, there are also different application methods of OTN in network construction, such as the whole OTN of WDM systems, OTN crossing-device applications in long-distance backbone and OTN crossing-device applications in MAN.

The introduction of OTN interfaces in WDM system can complete performance and fault monitoring for end to end wavelength path. OTN can realize the transparent transmission of all kinds of client signals, which is the prerequisite for the router to use the 10GE interface.[14] The OTN interface is introduced gradually into WDM system, preparing for the introduction of large-capacity equipment of OTN in the future.

With the development of long-distance IP network, IP traffic surges greatly, and the long-distance backbone core node is facing increasing volume of services. And in order to make more effective use of IP network resources, improve the utilization rate of relay circuit or network operation quality, it is necessary to use large capacity OTN crossover equipment in the long-distance backbone network. The use of large-capacity OTN equipment can realize the rapid opening of the wavelength path of large particles and improve the service response capability. After the ASON control plane loaded, a variety of protection and recovery methods based on ASON can be provided to improve the reliability of the backbone transmission network.[15]

The application of OTN in MAN is more complicated, and the corresponding technology is more competitive. In order to improve fiber utilization, OADM/ROADM based on wavelength level particle scheduling is an inevitable choice for WDM system construction in MAN or LAN construction. At present, the main domestic and foreign runners are very concerned about the development and application of OTN technology, and most of them have realized the OTN transmission interface function in WDM.

NEW WORDS AND PHRASES

undergo *vt*. 经历,经受
tremendous *a*. 极大的,巨大的,惊人的
video *n*. 视频,录像
put forward 提出,拿出
accommodate *v*. 容纳,适应
flexible *a*. 灵活的,有弹性的
schedule *v*. 安排,预定,调度
complete *vt*. 完成,使完满
maintenance *n*. 维护,维修
limitation *n*. 限制,极限
emphasize *vt*. 强调
mechanism *n*. 机制,原理
particle *n*. 颗粒,粒子
multi-wavelength 多波长
adopt *v*. 采取,接受
asynchronous *a*. 异步的,不同时的
mapping *n*. 映射,映像

multiplexing	多路复用,多工
configuration	n. 配置,结构
adaptability	n. 适应性,适应力
structure	n. 结构,构造
corresponding	a. 相当的,相应的
tandem	a. 串联的
fault	n. 故障,缺点
intelligent	a. 智能的,聪明的
survivability	n. 生存性,生存能力
adapter	n. 适配器,接合器
photoelectric	a. 光电的
hybrid	a. 混合的,杂种的
tributary	a. 附属的,辅助的
mainstream	n. 主流,主要倾向
external	a. 外部的,外用的
span	v. 跨越,持续
complement	vt. 补充,补足
transparent	a. 透明的,显然的
prerequisite	n. 先决条件
surge	vi. 汹涌,起大浪
utilization	n. 利用,使用
complicated	a. 难懂的,复杂的
competitive	a. 竞争的,比赛的
inevitable	a. 必然的,不可避免的

NOTES

1. 题目译成:光传送网 OTN(Optical Transmission Network)。OTN 是以波分复用技术为基础,在光层组织网络的传送网,是下一代骨干传送网。

2. carried by communication networks 过去分词短语做后置定语修饰 services。本句译文:近年来,通信网络所承载的业务发生了巨大的变化。

3. based on optical layer 过去分词短语做后置定语,修饰 WDM technology。to provide high capacity of transmission 不定式短语做后置定语,修饰 advantage。due to 由于。本句译文:基于光层的 WDM 技术,其多波长信道传输特性决定了它具有提供高传输容量的优点。

4. 本句是被动句。lacking of effective network maintenance and management tool 现在分词短语对句子前面部分做说明。本句译文:然而,目前的 WDM 网络主要应用于点对点模式,缺乏有效的网络维护和管理工具。

5. which can realize the transmission of large particles 为非限定性定语从句。本句

译文：OTN 在光层采用 WDM 技术，可以实现大颗粒的传送。

6. Compared with SDH VC-12/VC-4 scheduling particles 为过去分词短语；which can significantly improve the adaptability and transmission efficiency of high-bandwidth data customer services 是由 which 引导的非限定性定语从句。本句译文：相比于 SDH 的 VC-12/VC-4 的调度颗粒，OTN 复用、交叉和配置的颗粒明显大很多，能够显著提升高带宽数据客户业务的适配能力和传送效率。

7. 本句主语有两个谓语；providing advanced performance and fault monitoring 现在分词短语对句子前面部分进行说明。本句译文：OTN 技术可提供强大的 OAM 功能，并支持多达六个的串联连接监控(TCM)功能，提供高级性能和故障监控。

8. based on cross-functional ODU*k* 过去分词短语做后置定语，修饰 equipment。realizing flexible scheduling and protection to large-particles services 现在分词短语表示伴随情况。本句译文：基于交叉功能 ODU*k* 的 OTN 设备可以实现从 SDH-155M 到 2.5G / 10G / 40G 的电路交换颗粒，实现对大颗粒业务的灵活调度和保护。

9. improving the flexibility and survivability of network configuration 现在分词短语表示伴随情况。本句译文：OTN 设备还可以引入基于 ASON 的智能控制平面，提高网络配置的灵活性和生存性。

10. that supports the interfaces of OTN 是由 that 引导的定语从句，修饰 WDM equipment。which include the line interface and tributary interface 是由 which 引导的非限定性定语从句。本句译文：OTN 终端多路复用器是支持 OTN 接口的 WDM 设备，包括线路接口和支路接口(也称为业务接口或域间网络接口)。

11. similar to，类似于。providing flexible circuit scheduling and protection for OTN network 现在分词短语表示伴随情况。本句译文：与现有的 SDH 交叉连接设备类似，OTN 交叉设备完善了电路 ODU*k* 级交叉功能，为 OTN 网络提供灵活的电路调度和保护。

12. providing a variety of external interfaces and OTU*k* interfaces 现在分词短语表示伴随情况。本句译文：OTN 交叉设备可以独立存在，类似于 SDH DXC 设备，提供各种外部接口和 OTU*k* 接口。

13. that need to be scheduled 由 that 引导定语从句修饰 other services。本句译文：波长级服务直接跨越 OCh，其他需要调度的服务必须通过 ODU*k* 进行交叉。

14. which is the prerequisite for the router to use the 10GE interface 是由 which 引导的非限定性定语从句。本句译文：OTN 可以实现对各种客户端信号的透明传输，这是路由器使用 10GE 接口的前提条件。

15. 本句是被动句。based on ASON 过去分词短语做后置定语，修饰 methods。to improve the reliability of the backbone transmission network 不定式短语表示目的。本句译文：在 ASON 控制平面加载后，还可以提供多种基于 ASON 的保护和恢复方法，提高骨干传输网络的可靠性。

EXERCISES

一、请将下述词组译成英文

1. 快速发展的数据业务 2. IPTV 和视频发展 3. 提供高带宽
4. 快速调度的传输网络 5. 便利的网络维护和管理 6. 满足业务的需要
7. 使用 SDH 和 WDM 技术 8. 电层业务 9. 类似于 SDH DXC 设备
10. 提供有力的 OAM 能力 11. 基于光层的 WDM 技术 12. 传统 SDH 的优势
13. 大颗粒宽带业务

二、请将下述词组译成中文

1. the flexible scheduling, management and protection mechanism
2. the transmission characteristics of multi-wavelength channel
3. to realize the transmission of large particles
4. the asynchronous mapping and multiplexing at electrical layer
5. to improve the adaptability and transmission efficiency
6. a complete system structure of optical layer and electrical layer
7. the OTN equipments based on cross-functional ODU*k*
8. to introduce ASON-based intelligent control plane
9. to improve the flexibility and survivability of network configuration
10. the interface adapter and line interface processing
11. the mainstream manufacturers of WDM system
12. to provide a variety of external interfaces and OUT*k* interfaces
13. the photoelectric hybrid cross of OTN equipment
14. the different application methods of OTN
15. in order to make more effective use of IP network
16. the introduction of large-capacity equipment of OTN
17. the use of large-capacity OTN equipment
18. a variety of protection and recovery methods

三、选择合适的答案填空

1. Transmission networks must provide high bandwidth ________ this growth.

A. accommodate B. accommodation
C. accommodated D. to accommodate

2. The traditional SDH transmission network has the service scheduling capability of small particle and ________ transmission capacity.

A. limited B. limit
C. to limit D. limiting

3. OTN uses asynchronous mapping and multiplexing at the electrical layer ________ ODU*k*(*k*=1,2,3,4,flex) crossing-connection particles.

A. support B. supporting

C. to support D. supported

4. Electrical-crossing equipment of OTN can combine with OCh crossing equipment (ROADM or PXC) to provide scheduling function of ODU*k* electrical layer and OCh optical ________.

A. label B. layer

C. limit D. data

5. Typical equipment of this type is OSN6800 ________ by Huawei, which supports OCh-crossing based on ROADM and ODU1/GE crossing.

A. to produce B. producing

C. produced D. produce

6. The OTN interface is introduced gradually into WDM system, ________ for the introduction of large-capacity equipment of OTN in the future.

A. preparing B. to prepare

C. prepared D. prepare

四、根据课文内容选择答案

1. More importantly, it requires a transport network that can be quickly and flexibly scheduled, complete and convenient network maintenance and management (OAM functions) to meet ________.

A. the transmission network

B. the needs of the service

C. accommodate this growth

D. service of large particle

2. The traditional SDH transmission network has the service scheduling ________ and limited transmission capacity.

A. management and protection mechanism

B. broadband service of large particle

C. transmission characteristics of multi-wavelength

D. capability of small particle

3. Optical scheduling system (such as ROADM) can achieve ________ similar to SDH, but it is difficult to be applied to a wide range of networks due to the limitation of physical layer and wavelength.

A. the scheduling and protection ability

B. effective network maintenance and management tool

C. advantages of traditional SDH and WDM

D. management and protection mechanism

4. Compared with SDH VC-12/VC-4 scheduling particles, ________ multiplexing, crossover and configuration are significantly larger, which can significantly improve the adaptability and transmission efficiency of high-bandwidth data customer services.

A. advantages of traditional SDH

B. the limitation of physical layer

C. the transmission particles of OTN

D. broadband service of large particle

5. The OTN equipments can also introduce ASON-based ________, improving the flexibility and survivability of network configuration.

A. circuit switching particles

B. intelligent control plane

C. line interface processing

D. inter-domain network interface

6. There are several forms of OTN equipment, such as OTN ________, OTN electrical-crossing equipment, and OTN photoelectric hybrid-crossing equipment.

A. optical transmission section

B. end-to-end wavelength path

C. cross-connection equipment

D. terminal multiplexer

五、请将下列短文译成中文

1. In recent years, the services carried by communication networks have undergone tremendous changes. Data services develop rapidly, especially broadband services, such as IPTV and video development.

2. More importantly, it requires a transport network that can be quickly and flexibly scheduled, complete and convenient network maintenance and management (OAM functions) to meet the needs of the service. Currently the main transmission network is using SDH and WDM technology, but they both have some limitations.

3. SDH technology emphasis on electrical-layer service, it has flexible scheduling, management and protection mechanism and perfect OAM function. The traditional SDH transmission network has the service scheduling capability of small particle and limited transmission capacity. It cannot transmit efficiently broadband service of large particle.

4. However, the current WDM networks mainly are applied in point to point mode, lacking of effective network maintenance and management tool. Optical scheduling system (such as ROADM) can achieve the scheduling and protection ability similar to SDH, but it is difficult to be applied to a wide range of networks due to the limitation of physical layer and wavelength.

5. OTN technology has both of the advantages of traditional SDH and WDM. OTN adopts WDM technology in optical layer, which can realize the transmission of large particles. OTN uses asynchronous mapping and multiplexing at the electrical layer to support ODUk (k=1,2,3,4,flex) crossing-connection particles.

6. The OTN technology includes a complete system structure of optical layer and

electrical layer, each layer has a corresponding mechanism of network management, monitoring and survival. The OTN technology can provide powerful OAM capabilities, and enables up to six tandem connection monitoring (TCM) capabilities, providing advanced performance and fault monitoring.

7. The OTN equipment shall have the functions of customer interface, interface adapter and line interface processing. There are several forms of OTN equipment, such as OTN terminal multiplexer, OTN electrical-crossing equipment, and OTN photoelectric hybrid-crossing equipment.

8. With the development of long-distance IP network, IP traffic surges greatly, and the long-distance backbone core node is facing increasing volume of services. And in order to make more effective use of IP network resources, improve the utilization rate of relay circuit or network operation quality, it is necessary to use large capacity OTN crossover equipment in the long-distance backbone network.

参考译文

光传送网 OTN

近年来,通信网络所承载的业务经历了巨大的变化。数据业务发展迅速,尤其是宽带业务,如 IPTV 和视频开发。数据业务对传输网络提出了新的要求。传输网络必须提供高带宽以适应这种增长。更重要的是,它需要传输网络可以快速灵活地调度以满足业务的需要。目前主要的传输网络使用 SDH 和 WDM 技术,但它们都有一些局限性。

SDH 技术强调电层服务,具有灵活的调度、管理和保护机制以及完善的 OAM 功能。传统 SDH 传输网业务调度颗粒小,传送容量有限,不能有效地传送宽带大颗粒业务。基于光层的 WDM 技术,多波长信道的传输特性决定了它具有提供高传输容量的优点。但是,目前的 WDM 网络主要应用于点对点模式,缺乏有效的网络维护和管理工具。光调度系统(如 ROADM)可以实现类似于 SDH 的调度和保护能力,但由于物理层和波长的限制,很难应用于大范围的网络。

OTN 技术兼有传统 SDH 和 WDM 的优势。OTN 在光层采用 WDM 技术,可以实现大颗粒业务的传送。OTN 在电层使用异步映射和复用以支持 ODUk(k=1,2,3,4,flex)的交叉连接颗粒。相比于 SDH 的 VC-12/VC-4 的调度颗粒,OTN 复用、交叉和配置的颗粒明显大很多,能够显著改善高带宽数据客户业务的适配能力和传送效率。

OTN 技术包括完整的光层和电层系统结构,每层都有相应的网络管理、监控和生存机制。OTN 技术可提供强大的 OAM 功能,并支持多达六个的串联连接监控(TCM)功能,提供高级性能和故障监控。基于交叉功能 ODUk 的 OTN 设备可以实现从 SDH-155M 到 2.5G / 10G / 40G 的电路交换颗粒,实现对大颗粒业务的灵活调度和保护。OTN 设备还可以引入基于 ASON 的智能控制平面,提高网络配置的灵活性和生存性。

OTN 设备应具有客户接口、接口适配器以及线路接口处理功能。有几种形式的

OTN 设备,如 OTN 终端复用器、OTN 电交叉设备和 OTN 光电混合交叉设备。

OTN 终端多路复用器是支持 OTN 接口的 WDM 设备,包括线路接口和支路接口(也称为业务接口或域间网络接口),可实现端到端波长通道的性能和故障监控。目前主流 WDM 系统厂商基本采用 OTN 结构,并且在 G.709 标准中支持 OTN 接口,因此它们都是 OTN 终端多路复用器。

与现有的 SDH 交叉连接设备类似,OTN 交叉设备完善了电路 ODU*k* 级交叉功能,为 OTN 网络提供灵活的电路调度和保护。OTN 电交叉设备可以独立存在,类似于 SDH DXC 设备,提供各种外部接口和 OTU*k* 接口。它们还可以与 OTN 终端多路复用器集成,提供支持 WDM 传输的光复用段和光传输段功能。

OTN 的电交叉设备可以与 OCh 交叉设备(ROADM 或 PXC)结合以提供 ODU*k* 电层和 OCh 光层调度功能。波长级业务直接跨越 OCh,其他需要调度的业务必须通过 ODU*k* 进行交叉。两者的结合可以相互补充,也可以避免它们的弱点。这种大容量调度设备是 OTN 设备的光电混合交叉。

这类设备的典型产品是华为生产的 OSN6800,它支持基于 ROADM 和 ODU1 / GE 交叉的 OCh 交叉。ODU1 交叉容量为 320 GB 并支持 ASON 控制平面。其他制造商,如中兴通讯,也在开发类似的产品。

根据 OTN 设备的不同存在形式,OTN 在网络建设中的应用方式也不同,如 WDM 系统的完全 OTN 应用,在长途骨干网中的 OTN 交叉设备应用和在城域网中的 OTN 交叉设备应用。

在 WDM 系统中引入 OTN 接口可以完成端到端波长路径的性能和故障监控。OTN 可以实现对各种客户端信号的透明传输,这是路由器使用 10GE 接口的前提条件。逐步在 WDM 系统中引入 OTN 接口,为将来引入 OTN 的大容量设备做好准备。

随着远程 IP 网络的发展,IP 业务量激增,远程骨干核心节点面临着日益增长的业务量。并且为了更有效地利用 IP 网络资源,提高中继电路的利用率或网络运营质量,有必要在远程骨干网中使用大容量 OTN 交叉设备。使用大容量 OTN 设备,可以实现大颗粒波长路径的快速开放,改善服务响应能力。在 ASON 控制平面加载后,还可以提供多种基于 ASON 的保护和恢复方法,以提高骨干传输网络的可靠性。

在城域网 MAN 中使用 OTN 情况比较复杂,相应的技术更具竞争力。为了提高光纤利用率,在城域网或局域网建设中建设 WDM 系统,基于波长级颗粒调度的 OADM / ROADM 是必然的选择。目前国内外主要的运营商都非常关注 OTN 技术的发展和应用,大多数运营商已经在 WDM 中实现了 OTN 传输接口功能。

UNIT 2

PASSAGE

Synchronous Digital Hierarchy[1]

SDH is an international standard for high-speed synchronous optical telecommunication networks—a synchronous digital hierarchy.

Work started on SDH standards in CCITT's Study Group XVIII in June 1986. The objective was to produce a worldwide standard for synchronous transmission systems which provides network operators with a flexible and economic network.

In November 1988 the first SDH standards were approved—G. 707, G. 708 and G. 709. These standards defined transmission rates, signal format, multiplexing structures and tributary mappings for the network node interface (NNI)—the international standard interface for synchronous digital hierarchy.

In addition to defining standards covering the NNI, CCITT also embarked on series of standards governing the operation of synchronous multiplexers (G. 781, G. 782 and G. 783) and SDH network management[2] (G. 784). It is the standardization of these aspects of SDH equipment that will deliver the flexibility required by network operators to cost-effectively manage the growth in bandwidth and provisioning of new customer services expected in the next decade.[3]

The SDH standards are based on the principles of direct synchronous multiplexing which is the key to cost-effective and flexible telecommunication networking. In essence, it means that individual tributary signals may be multiplexed directly into a higher rate SDH signal without intermediate stages of multiplexing. SDH network elements can then be interconnected directly with obvious cost and equipment savings over the existing network.

Advanced network management and maintenance capabilities are required to effectively manage the flexibility provided by SDH. Approximately 5% of the SDH signal structure is allocated to supporting advanced network management procedures and practices.

The SDH signal is capable of transporting[4] all the common tributary signals found in today's telecommunication networks. This means that SDH can be deployed as an overlay to the existing signal types. In addition, SDH has the flexibility to readily accommodate new types of customer service signals that network operators will wish to support in the future.

SDH can be used in all traditional telecommunications application areas. SDH therefore makes it possible for a unified telecommunication network infrastructure to evolve. The fact that SDH provides a single common standard for this telecommunications network means that equipment supplied by different manufacturers may be interconnected directly.

Now, let's take a look at the network "building blocks" and how they are configured. These network elements are now all defined in CCITT standards and provide multiplexing or switching functions.

Line terminal multiplexer (LTM): LTM can accept a number of tributary signals and multiplex them to the appropriate optical SDH rate carrier, i.e. STM-1, STM-4 or STM-16. The input tributaries can either be the existing PDH signals such as 2, 34 and 140 Mbit/s or lower rate SDH signals. LTM forms the main gateway from the PDH to the SDH.

Add-drop multiplexer (ADM): a particular type of terminal multiplexer designed to operate in a through mode fashion.[5] Within the ADM it is possible to add channels to, or drop channels from the "through" signal. ADM is generally available at the STM-1 and STM-4 interface rates and may add/drop a variety of tributary signals, i.e. 2, 34 or 140 Mbit/s.

The ADM function is one of the major advantages resulting from the SDH since the similar function within a PDH network, required banks of hardwired back-back terminals.

Synchronous DXC: these devices will form the cornerstone of the new synchronous digital hierarchy. They can function as semi-permanent switches for transmission channels and can switch at any level from 64 kbit/s up to STM-1. Generally such devices have interfaces at STM-1 or STM-4. The DXC can be rapidly reconfigured, under software control, to provide digital leased lines and other services of varying bandwidth.

For clarity, a single frame in the STM-1 can be represented by a 2-dimensional map (see Fig. 2-1). The 2-dimensional map comprises 9 rows and 270 columns of boxes. Each box represents a single 8-bits byte within the synchronous signal. Six framing bytes appear in the top left corner of the 2-dimensional map. These framing bytes act as a marker, allowing[6] any byte in the frame to be easily located.

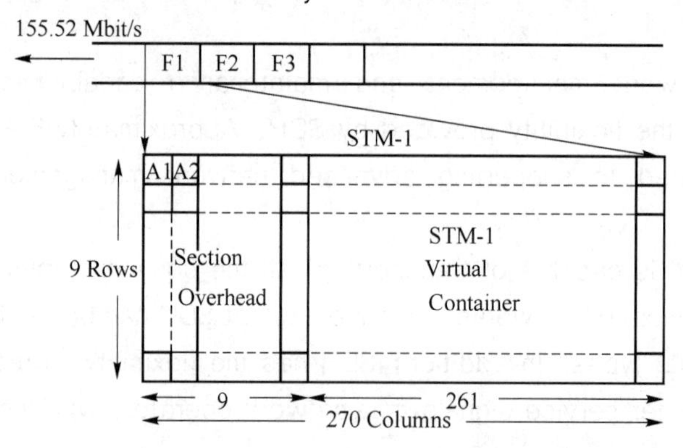

Fig. 2-1 Synchronous Transport Frame for STM-1

The signal bits are transmitted in a sequence starting with those in the first row. The order of transmission is from left to right. After transmission of the last byte in the frame (the byte located in row 9, column 270), the whole sequence repeats starting with the 6 framing bytes of the following frame.

A synchronous transport frame comprises two distinct and readily accessible parts within the frame structure—a virtual container part and a section overhead part. Virtual container (VC) is arranged mainly for user's information to be transmitted through the network, but the section overhead (SOH) provides the facilities required to support and maintain the transportation of a VC between nodes in a synchronous network, such as alarming monitoring, bit-error monitoring and data communication channels.

NEW WORDS AND PHRASES

standard	*n*. 标准
hierarchy	*n*. 体系，分层结构
synchronous	*a*. 同步的
approve	*v*. 批准，审订，通过
tributary	*a*. 支流的，从属的，辅助的
map	*vi*. 绘制，设计，映射，变换
node	*n*. 节点，结
embark	*vi*. 从事，开始做
provision	*n*. 准备，预备
essence	*n*. 本质，精华，核心
in essence	本质上，实质上
advanced	*a*.先进的，前进的，高深的
maintenance	*n*. 维护，保养
approximately	*ad*.近似地，大约地
procedure	*n*. 过程，步骤，程序，手续
overlay	*n*. 覆盖层，涂盖层
accommodate	*v*. 供应，提供，容纳
unify	*v*. 使一致，使一体
configure	*vt*. 使成形，使具形体，构成
gateway	*n*. 门口，入口，通道
mode	*n*. 方式，样式，模式
variety	*n*. 变化，多样化，各种各样
cornerstone	*n*. 基石，柱石，基础
permanent	*a*. 永久的，持久的

lease	v. 出租
virtual	a. 虚的,假想的,虚拟的
container	n. 容器
section	n. 节,段,部分
dimensional	a. 维的,度的
marker	n. 记号,符号,标志
intact	ad.(a.)未经触动地(的),完整地(的),未受损地(的)
capacity	n. 容量,能力,容纳量
distinct	a. 不同的,有区别的,清晰的,明显的
accessible	a. 很快的,容易的,简便的,直接的
overhead	a. 可接入的
assemble	v. 集合,装配,聚集

NOTES

1. 本篇课文涉及同步数字系列这一通信新技术,题目可译为:同步数字系列。

2. 本句中,defining 为动名词,做介词宾语;covering 为现在分词,作为 standard 的定语, governing 引起的现在分词短语亦做定语,修饰 a series of standards。因而全句可译成:除定义了 NNI 有关的标准外,CCITT 还着手制定了决定着同步复用设备的运行以及 SDH 网络管理的一系列标准。

3. 这是一个强调句。强调句的基本形式为:

It is (was)+强调部分+that(which)+谓语及其他成分

强调句一般可译成:正是……。另外,本句的 to cost-effectively manage the growth in bandwidth and provisioning of new customer services expected in the next decade 是一个动词(manage)不定式短语,在句中做目的状语。

全句可译成:正是由于 SDH 设备在这些方面的标准化,才提供了网络运营者所期望的灵活性,以便能低价高效地应付带宽方面的增长并为后十年中将出现的新的用户业务做好准备。

4. be capable of doing… 能够做……。

5. add-drop multiplexer 中的 add-drop 意为"上下话路",可译为"分插复用单元"。designed 为过去分词,做定语,修饰 a particular type of terminal multiplexer。本句可译成:分插复用单元——一种特殊类型的终端复用单元,它是以"贯通"模式运行的。

6. 句中的 allowing 为现在分词,做状语,用以表示目的或结果。全句可译成:这些成帧字节起着标志的作用,它使帧中的任何字节极易被确定位置。

EXERCISES

一、请将下述词组译成英文

1. 同步数字系列　　2. 国际标准　　3. 信号格式

4. 网络节点接口
5. 支路信号
6. 数字交叉连接
7. 网络管理
8. 网络维护
9. 网络运营者
10. 传输速率
11. 支路映射
12. 灵活性
13. 用户业务
14. 覆盖层
15. 制造商
16. 同步传输帧
17. 线路终端复用器
18. 分插复用器
19. 再生中继器
20. 灵敏度
21. 虚容器
22. 成帧字节
23. 段开销
24. 端到端传输
25. 误码监视
26. 信号处理节点
27. 净负荷
28. 指针

二、请将下述词组译成中文

1. synchronous transmission system
2. the standard covering the NNI
3. the international standard interface
4. direct synchronous multiplexing
5. flexible telecommunication networking
6. point-to-point transmission technology
7. advanced network management
8. the equipment supplied by different manufacturers
9. the flexibility provided by SDH
10. operator of synchronous multiplexers
11. telecommunication networking
12. tributary signals
13. maintenance capabilities
14. unified telecommunication network infrastructure
15. building blocks
16. terminal multiplexer
17. through-mode fashion
18. synchronous DXC
19. varying bandwidth
20. individual tributary signals
21. transport system
22. optical carrier
23. 2-dimensional map
24. the order of transmission
25. framing byte
26. virtual container
27. section overhead
28. bit-error monitoring

三、选择合适的答案填空

1. In addition to defining standards ________ the NNI, CCITT also embarked on a series of standards ________ the operation of synchronous multiplexers and SDH Network Management.

A. cover, govern　　B. to cover, to govern

C. covering, governing　　D. covered, governed

2. Approximately 5% of the SDH signal structure is ________ to supporting ________ network management procedures and practices.

A. allocated, advanced　　B. allocate, advance

C. allocating, advancing　　D. to allocate, to advance

3. Add-drop multiplexer is a particular type of terminal multiplexer ________ in a through mode fashion.

A. designing operate　　B. design operating

C. designed operate　　D. designed to operate

4. The ADM function is one of the major advantages ________ from the SDH since the similar function within a PDH network, ________ banks of hardwired back-back terminals.

A. resulting, required　　B. result, require

C. resulting, requiring　　D. resulted, to require

5. The DXC can be rapidly reconfigured, under software control, ________ digital leased lines and other services of ________ bandwidth.

A. provided, varied　　B. provision, variation

C. providing, to vary　　D. to provide, varying

6. Six framing bytes appear in the top left corner of the 2-dimentional map. These framing bytes act as a marker, ________ any byte in the frame ________.

A. allow, be easily located　　B. to allow, to be easily locate

C. to allow, is easily located　　D. allowing, to be easily located

7. More importantly, however, signal capacity ________ within a synchronous transport frame ________ network transportation capabilities.

A. set aside, support　　B. is set aside, to support

C. is aside, supported　　D. to set, be supported

8. After ________ of the last byte in the frame, the whole sequence repeats with the 6 framing bytes of the following frame.

A. transmit, start　　B. transmitted, start

C. transmission, starting　　D. to transmit, to start

四、根据课文内容选择正确答案

1. The SDH signal is capable of transporting ________ found in today's telecommunication networks.

A. all the analog signals

B. all the common tributary signals

C. the voice signals

D. all the video signals

2. The fact that SDH provides a single common standard for the telecommunications network means that equipment supplied by different manufacturers ________.

A. can not be linked

B. could by very expensive

C. may not be produced

D. may be interconnected directly

3. Line terminal multiplexers can accept a number of ________ and multiplex them to appropriate optical SDH rate carrier, i.e. STM-1, STM-4 or STM-16.

A. signals from a satellite

B. voice and video signals from telecommunications network

C. tributary signals

D. signals from the optical fiber

4. The add drop mux (ADM) is the basic SDH building block for ________.

A. telephone switching

B. monitoring operations

C. alarm reporting

D. local access to synchronous networks

5. A synchronous transport frame comprises two distinct and readily accessible parts within the frame structure: ________.

A. framing bytes and scrambling bytes

B. a alarm monitoring part and a bit-error monitoring part

C. a virtual container part and a section overhead part

D. a SDH part and a PDH part

6. The Virtual Container is used ________ across the synchronous network.

A. transport a frame

B. to transmit the SDH signals

C. to provide the facilities, such as bit-error monitoring

D. to transport a tributary signal

7. Individual tributary signals are arranged ________ for end-to-end transmission across the SDH network.

A. out the virtual container

B. before the virtual container

C. within the virtual container

D. after the virtual container

8. Some signal capacity is allocated in each transport frame for "section overhead", this provides the facilities required ________ between nodes in an synchronous network.

A. to support the transmission of the data

B. to support the transmission of a VC

C. to ensure the clock can be recovered

D. to ensure there are tributary signals

五、请将下述短文译成中文

1. In November 1988 the first SDH standards were approved. These standards define transmission rates, signal format, multiplexing structures and tributary mappings for the network node interface (NNI)—the international standard interface for the synchronous digital hierarchy. It is the standardization of these aspects of SDH equipment that will deliver the flexibility required by network operators to cost-effectively manage the growth in bandwidth and provisioning of new customer services expected in the next decade.

2. The SDH standards are based on the principle of direct synchronous multiplexing which is the key to cost effective and flexible telecommunication networking. In essence, it means that individual tributary signals may be multiplexing directly into a higher rate SDH signal without intermediate stage of multiplexing. SDH Network Elements can then be interconnected directly with obvious cost and equipment savings over the existing networks.

3. The SDH signal is capable of transporting all the common tributary signals found in today's telecommunication networks. This means that SDH can be deployed as an overlay to the existing signal types. In addition, SDH has the flexibility to readily accommodate new types of customer service signals that network operators will wish to support in the future.

4. The restrictions of PDH are mainly reflected in the following:

- The incompatibility of the 1.5 Mbit/s and the 2 Mbit/s PDH hierarchies is detrimental to the development of international communications.
- The inflexibility of asynchronous multiplexing of stepped code speed adjustment and adding and dropping circuits restricts the move towards higher orders and further utilization of the enormous transmission capability of optical fiber.
- There are few overhead bits in the PDH frame structure, resulting in an inadequate OAM function.
- The PDH optical interfaces, including line codes, have not been standardized so that equipment produced by different manufacturers cannot interwork with each other and the line systems lack transverse compatibility.

5. The line terminal multiplexers can accept a number of tributary signals and multiplex them to the appropriate optical SDH rate carrier, for example, the STM-16.

The input tributaries can either be existing PDH signals or lower rate SDH signals. LTMs form the main gateway from the PDH network to the SDH.

6. Add-drop multiplexer is a particular type of terminal multiplexer designed to operate in a through mode fashion. Within the ADM it is possible to add channels to, or drop channels from the "through" signal. The ADM function is one of the major advantages resulting from the SDH since the similar function within a PDH network, required banks of hardwired back-back terminals.

7. Synchronous DXC will form the cornerstone of the new synchronous digital hierarchy. They can function as semi-permanent switches for transmission channels and can switch at any level from 64 kbit/s up to STM-1. Generally such devices have interfaces at STM-1 or STM-4. The DXC can be rapidly reconfigured, under software control, to provide digital leased lines and other services of varying bandwidth.

8. The advantages of SDH are mainly reflected in the following:

- Lower network element costs: With a common standard, compatible equipment will be available from many vendors. In a highly competitive market prices will be very attractive.
- Better network management: With better network management, operators will be able to more efficiently use the network and provide better service. The concept of TMN (Telecom Management Networks) is under study by CCITT. Some TMN standards defining management system interfaces already exist.
- Faster provisioning: If new circuits can be software defined to use existing spare bandwidth then provisioning will be much faster. The only new connection needed will be from the customer's premises to the nearest network access node.

9. The advantages of SDH are mainly reflected in the following:

- Better network utilization: With total control of routing customer circuits can be "groomed" or "hubbed" to make best use of network resources. Typically, all speech carrying circuits may be split from data circuits and routed for minimum delay. Data circuits, depending on the type, may be "hubbed" to a particular network DXC with the level of cross-connect needed.
- Better network survivability: With "real-time" rerouting possible, the network's operation support system will be able to take care of failure by simply reprogramming circuit paths. The built-in protection and reporting systems will automatically take care of simple transmission failures.
- Simpler handover: If all networks use equipment conforming to the same standard, the handover of circuits at the "network node interface" should be trouble free.
- Support of future services: Looking to the future, the SDH design will cater for

new services like high definition TV, wide area network backbone networks, broadband ISDN and new bandwidth-on-demand services. As the SDH operator will have total control of bandwidth allocation, any new service will be simple to provision.

10. For clarity, a single frame in the serial signal stream can be represented by a 2-dimentional map. The 2-dimentional map comprises 9 rows and 270 columns of boxes. Each box represents a single 8 bit byte within the synchronous signal. Six framing bytes appear in the top left corner of the 2-dimentional map. These framing bytes act as a marker, allowing any bytes in the frame to be easily located.

11. A synchronous transport frame comprises two distinct and readily accessible parts within the frame structure—a virtual container part and a section overhead part. Individual tributary signals are arranged within the virtual container for end-to-end transmission across the SDH network. The section overhead provides the facilities required to support and maintain the transportation of a VC between nodes in a synchronous network.

12. The virtual container is used to transport a tributary signal across the synchronous network. In most cases, this signal is assembled at the point of entry to the synchronous network and disassembled at the point of exit. Within the synchronous network, the virtual container is passed on intact between transport systems on its route through the network. However, the section overhead pertains only to an individual transport system and supports the transportation of the VC over that transport system. It is not transferred with the VC between transport systems.

13. To take care of small differences in the synchronous network, and simplify multiplexing and cross-connection of signals, the VC-4 is allowed to float within the payload capacity provided by the STM-1 frames. This means that the VC-4 may begin anywhere in the STM-1 payload capacity and is unlikely to be wholly contained in one frame. More likely than not, the VC-4 begins in one frame and ends in the next.

参考译文

同步数字系列

SDH,即同步数字系列,是为高速同步光纤通信网指定的一个国际标准。

CCITT 的第十八研究组早在 1986 年 6 月就开始了对 SDH 标准的研究工作。其目的在当时是要制定一个同步传输系统的世界性标准,而该标准将向网络运营者提供一个灵活的、经济的网络。

1988 年 11 月,第一批 SDH 标准——G. 707、G. 708 和 G. 709 获准通过。这些标准定义了网络节点接口(NNI)的传输速度、信号格式、复用结构和支路映射。网络节点接口,即同步数字系列的国际标准接口。

除定义了NNI有关的标准外,CCITT还着手制定了决定着同步复用设备的运行(G.781、G.782和G.783)以及SDH网络管理(G.784)的一系列标准。正是由于SDH设备在这些方面的标准化,才提供了网络运营者所期望的灵活性,从而能低价高效地应付带宽方面的增长并为后十年中将出现的新的用户业务做好准备。

SDH标准的基础是直接同步复用。这一原理是低价高效和灵活组网的关键。从本质上讲,这意味着各支路信号可以被直接复用到更高速率的SDH信号之中,而用不着中间级的复用阶段。诸SDH网络单元可以直接相互连接,因而显然比现在的网络更省钱、更省设备。

为了有效地发挥SDH提供的灵活性,需要先进的网络管理和维护能力。SDH信号结构中有近5%被用于支持先进的网络管理过程与实践。

SDH能够传输当今电信网中所有常见的支路信号。这意味着我们可用SDH作为现有信号类型的总包层。此外,对于网络运营者将来欲支持的各种新型的用户业务信号,SDH亦具有直接处理的灵活性。

SDH可应用于所有传统的电信应用领域。因此,SDH有可能使电信网的结构演变成为一个统一的网络。SDH为这个网络提供了单一的、通用的标准,这就意味着由不同厂家供应的设备可以直接互联起来。

现在,就让我们来看看网络的"组件"以及它们的构成方式。这些网络单元均已由CCITT的标准所定义,并提供复用或交换的功能。

线路终端复用单元(LTM):LTM可接受一批支路信号并将它们复用至适当的SDH速率的光载体上,即STM-1或STM-16。输入的支路信号可以是现有的PDH信号,例如2 Mbit/s、34 Mbit/s,也可以是较低速率的SDH信号。线路终端复用单元(LTM)形成了由PDH网络到SDH的主入口。

分插复用单元(ADM):一种特殊类型的终端复用单元,它是以"贯通"模式运行的。在分插复用单元(ADM)中,可以从"贯通"信号中上、下话路。ADM通常在STM-1和STM-4的接口速率下工作,并能上、下多种支路信号,即2 Mbit/s、34 Mbit/s或140 Mbit/s的信号。

ADM的功能是SDH的主要优势之一,因为在PDH网络中,完成类似功能需要一套硬布线的背对背的终端设备。

同步数字交叉连接单位(DXC):这些单元将成为新的同步数字系列的基石。它们能对传输信道起到半永久交换的作用,并可在从64 kbit/s直至STM-1速率的任何级别上进行交换。一般地,这种单元具有STM-1或STM-4的接口。DXC可以在软件控制下迅速地重构电路,以提供数字租用线路或变带宽的其他业务。

为清晰起见,在STM-1中的一帧可用一个二维的图形表示(见图2-1,第28页)。该二维结构由9行、270列个方框组成。每个方框代表同步信号中的一个8 bit字节。在二维图的左上角有6个成帧字节,这些成帧字节起着标志作用,它使帧中的任何字节极易被确定位置。

信号比特按这样的次序传送:首先传送第一行,传输次序由左到右。当帧中的最后一个字节(即位于第 9 行第 270 列的字节)被传送之后,就再重复整个次序——又从下一帧的 6 个成帧字节开始。

一个同步传输帧是由两个不同的,并可直接接入的两部分——虚容器部分和段开销部分组成。虚容器主要用于通过网络来传送用户信息,而段开销提供了支持和维护同步网络节点间传输 VC 所需要的开销,例如告警监视、误码检测和数据通信信道。

UNIT 3

PASSAGE

The Principle of PCM[1]

PCM is dependent on[2] three separate operations, sampling, quantizing, and coding. Many different schemes for performing[3] these three functions have evolved during recent years, and we shall describe the main ones. In these descriptions we shall see how a speech channel of telephone quality may be conveyed as a series of amplitude values, each value being represented[4], that is, coded, as a sequence of 8 binary digits.[5] Furthermore, we shall prove that a minimum theoretical sampling frequency of order 6.8 kilohertz (kHz) is required to convey a voice channel occupying[6] the range 300 Hz to 3.4 kHz. Practical equipments, however, normally use a sampling rate of 8 kHz, and if 8-digits per sample value[7] are used, the voice channel becomes represented by a stream of pulses with a repetition rate of 64 kHz. Fig.3-1 illustrates the sampling, quantizing, and coding processes.

Reexamination of our simple example shows us that the speech signal of maximum frequency 3.4 kHz has been represented by a signal of frequency 64 kHz. However, if only 4-digits per sample value had been used, the quality of transmission would drop, and the repetition rate of the pulses would be reduced to 32 kHz. Thus the quality of transmission is dependent on the pulse repetition rate, and for digital communication systems these two variables may be interchanged most efficiently.[8]

Digital transmission provides a powerful method for overcoming noisy environments. Noise can be introduced into transmission path in many different ways: perhaps via a nearby lightning strike, the sparking of a car ignition system, or the thermal low-level noise within the communication equipment itself.[9] It is the relationship of the true signal to the noise signal, known as the signal-to-noise ratio[10], which is of the most interest to the communication engineer. Basically, if the signal is very large compared to the noise level[11], then a perfect message can take place; however, this is not always the case. For example, the signal received from a satellite, located in far outer space[12], is very weak and is at a level only slightly above that of the noise. Alternative examples may be found within terrestrial systems where, although the message signal is strong, so is the noise power.

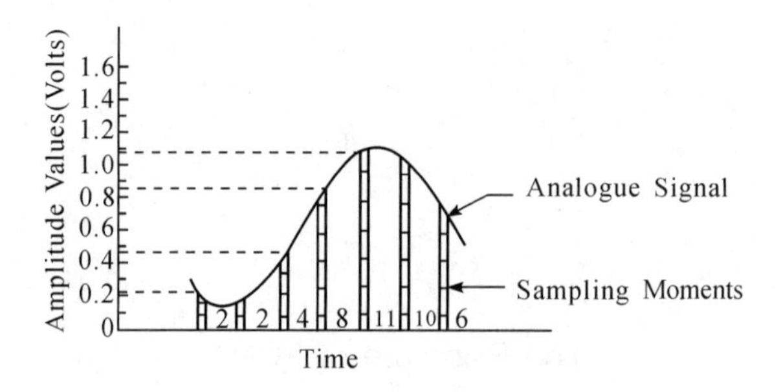

Amplitude Value	Binary Coded Equivalent	Pulse Code Modulated Signal
1	0000	
2	0001	
3	0010	
4	0011	
5	0100	
6	0101	
7	0110	
8	0111	
9	1000	
10	1001	
11	1010	
12	1011	
13	1100	
14	1101	
15	1110	
16	1111	

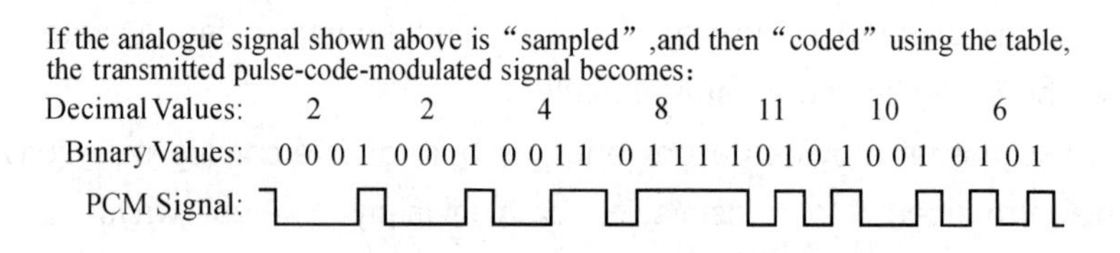

Fig. 3-1 The Sampling and Coding Processes, and the Resultant PCM Signal

If we consider binary transmission, the complete information about a particular message will always be obtained by simply detecting the presence or absence of the pulse. By comparison, most other forms of transmission systems convey the message information using the shape, or level of the transmitted signal; parameters that are most easily affected by the noise and attenuation introduced by the transmission path. [13] Consequently there is an inherent advantage for overcoming noisy environments by choosing digital transmission.

So far in this discussion we have assumed that each voice channel has a separate coder, the unit that converts sampled amplitude values to a set of pulses; and decoder, the unit that performs the reverse operation. This need not be so, and systems are in operation where a single codec (i.e., coder and its associated decoder) is shared between 24, 30, or even 120 separate channels. A high-speed electronic switch is used to present the analog information signal of each channel, taken in turn[14], to the codec. The codec is then arranged to sequentially sample the amplitude value, and code this value into the 8-digit sequence. Thus the output to the codec may be seen as a sequence of 8 pulses relating to channel 1, and then channel 2, and so on. This unit is called a time division multiplexer (TDM), and is illustrated in Fig. 3-2. The used multiplexing principle is known as word interleaving, since the words, or 8-digit sequences, are interleaved in time.

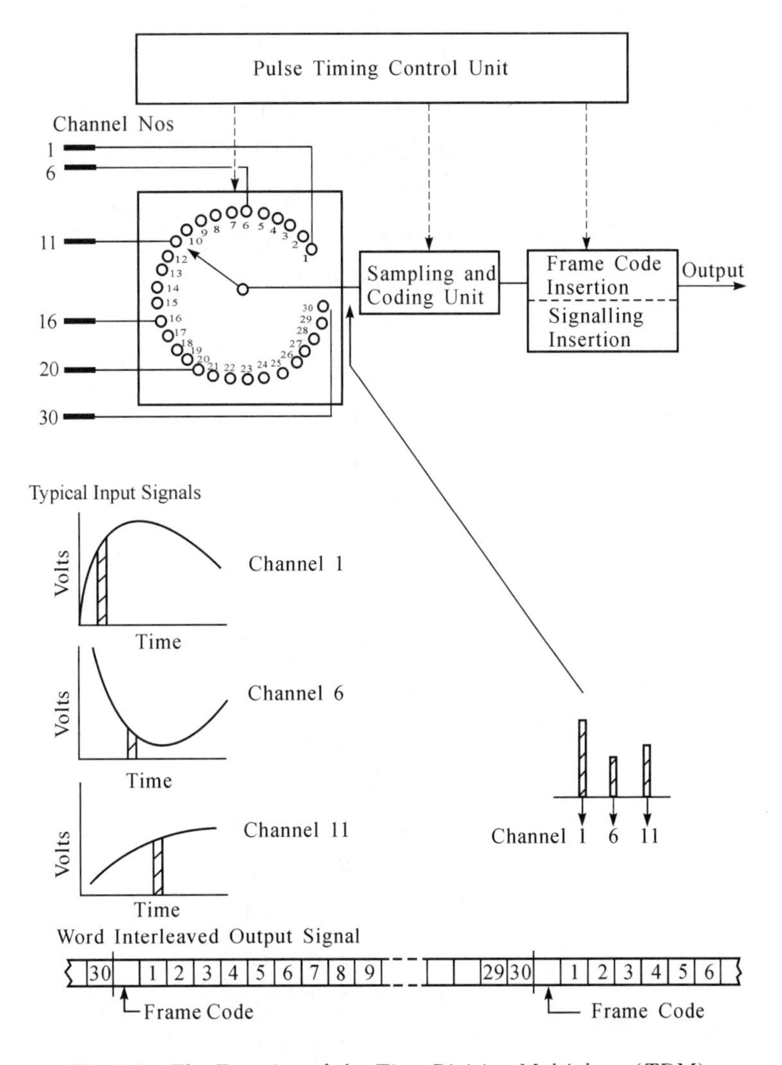

Fig. 3-2 The Function of the Time Division Multiplexer (TDM)

At the receiving terminal a demultiplexer is arranged to separate the 8-digit sequences into the appropriate channels. The reader may ask, how does the demultiplexer know which group of 8-digits relates to channel 1, 2, and so on? Clearly this is important! The problem is easily overcome by specifying a frame format, where at the start of each frame a unique sequence of pulses called the frame code, or synchronization word, is placed so as to[15] identify the start of the frame. A circuit of the demultiplexer is arranged to detect the synchronization word, and thereby it knows that the next group of 8-digits corresponds to channel 1. The synchronization word reoccurs once again after the last channel has been received.[16]

NEW WORDS AND PHRASES

principle	*n*. 原理
be dependent on	依赖,取决于
sample	*vt*. 采样;*n*. 样值
quantize	*vt*. 量化,分层
code	*vt*. 编码;*n*. 码
scheme	*n*. 方案,设计,安排
describe	*vt*. 叙述,描述
description	*n*. 叙述,描述
amplitude	*n*. 幅,幅度
binary	*a*. 二进制的
minimum	*n*. 最小值,最小量
theoretical	*a*. 理论上的
repetition	*n*. 重复,反复
reexamination	*n*. 再审查,重考
maximum	*n*. 最大值
reduce	*v*. 减少,缩小
interchange	*v*. 互换,转换, 相互影响
method	*n*. 方式,方法,手段
overcome	*v*. 克服,打败,征服
environment	*n*. 环境,周围情况
lightning	*n*. 电光,闪电,雷电
strike	*v*. (*n*.) 击,敲,打
spark	*vi*. 发火花,打火,闪光
ignition	*n*. 点火,点火装置
signal-to-noise ratio	信噪比

satellite	*n*. 卫星
terrestrial	*a*. 地球的，地面的，大地的
by comparison	比较起来，相对之下
parameter	*n*. 参数，系数
attenuation	*n*. 衰减，衰耗
inherent	*a*. 固有的，内在的
assume	*v*. 假设，假定
decoder	*n*. 解(译)码器
codec	*n*. 编译码器
interleave	*vt*. 交插，交错，插接
appropriate	*a*. 适当的，合适的
unique	*a*. 唯一的，独特的
reoccur	*vi*. 再发生，再次发生

NOTES

1. 本篇课文涉及数字通信领域，题目为：PCM 原理。

2. be dependent on，依靠，依赖，取决于。

3. performing 是 perform 的动名词。动名词虽为名词，但仍保留着动词的某些特征，例如它仍可带有动词宾语，仍可由副词进行修饰。本句中的 these three functions 就是 performing 的动词宾语。

本书中，这类例子相当普遍。例如第四段中的最后一句：

Consequently there is an inherent advantage for overcoming noisy environments by choosing digital transmission.

句中，overcoming 和 choosing 都是动名词且带有自己的动词宾语。

4. each value being represented 这是一种独立分词结构，用来表示一种伴随状态，做状语。该句可译为：而每一幅值可被表示为……。独立分词结构在科技英语文章中相当常见，例如：

There are many kinds of steel, each having its uses in industry.

译为：钢有许多种，在工业中每种都有它自己的用途。

独立分词结构还可以表示时间、原因或条件等，对主句进行补充说明。例如表示原因：

The resistance being very high, the current in the circuit was low.

译为：由于电阻很高，故电路里的电流很小。

5. as a sequence of 8 binary digits 句中的 as 意为“作为，当作，成为”。该短语可译为：(作为)8 位二进制码的序列。

6. occupying 是 occupy 的现在分词。动词的-ing 形式可以作为名词用，称为动名词,本课注释 3 已对此做了介绍;动词的-ing 形式亦可作形容词用(称现在分词),它一般修饰该动作的发出者,而且它亦可带有自己的宾语。例如本句中的“a voice channel occupying the range 300 Hz to 3.4 kHz”,该短语可译为:占有 300 Hz 到 3.4 kHz 频率范围的话路。

7. 8-digits per sample value 可译为:每样值 8 位码。

8. most efficiently,极有效地,极明显地。

9. 本句中的几个词组可译为:

lightning strike,打雷;

sparking of a car ignition system,汽车点火系统的打火;

thermal low-level noise,低电平的热噪声。

10. known as the signal-to-noise ratio 译为:称为信号噪声比。

11. compared to the noise level 译为:与噪声电平相比。

12. located in far outer space 这是一种分词短语结构。located 是动词的-ed 形式,称为过去分词,起形容词的作用,在句中做定语。这种分词往往用来修饰原来动词动作的承受者,例如:

The trees planted by me have grown up.

译为:我种的树已长大了。

再例如:

Electromotive force results in electrical pressure, compared to water pressure.

译为:电动势产生电压,电压好比水压。

所以课文中的这一短语可译为:位于遥远太空中的(卫星)。

13. 本句中,需要说明下述问题。

by comparison,相比之下。

using the shape, or level of transmitted signal 为现在分词短语结构,做状语,见本书的“主要语法现象”,可译为:用传输信号的波形或电平(来传送信息)。

类似的例子还有:

The rockets rose hissing over the launching site. (火箭在发射场的上空嘶嘶地上升。)

再例如:

They sat together, carefully studying the design of the circuits. (他们坐在一起仔细地研究着那些电路的设计。)

对课文中的这个句子,全句可译为:相比之下,许多其他形式的传输系统是利用被传信号的波形或电平来传送信息的,而这些参数又极易受到传输途径中噪声和衰耗的影响。

14. taken in turn,轮流地,依次地。

15. so as to,以便,为了。

16. 本句可译为:当收到最后一路之后,同步码字又再次出现。

EXERCISES

一、请将下述词组译成英文

1. 抽样、量化与编码　　2. 话路　　3. 幅值　　4. 抽样频率
5. 抽样速率　　6. 脉冲流　　7. 重复率　　8. 编码过程
9. 模拟信号　　10. 传输质量　　11. 数字通信　　12. 数字传输
13. 含噪声的环境　　14. 传输路由　　15. 信噪比　　16. 信号电平
17. 地面系统　　18. 噪声功率　　19. 二进制传输　　20. 反向操作
21. 8位码序列　　22. 接收端　　23. 帧格式　　24. 同步字

二、请将下述词组译成中文

1. the schemes for performing these three functions
2. a series of amplitude values
3. a speech channel of telephone quality
4. a sequence of 8-binary digits
5. a minimum theoretical sampling frequency
6. a voice channel occupying the range 300 Hz to 3.4 kHz
7. 8-digits per sample value
8. the sparking of a car ignition system
9. the stream of the pulses with a repetition rate of 64 kHz
10. the relationship of the true signal to the noise signal
11. the signal received from a satellite
12. the complete information about a particular message
13. the shape of the transmitted signal
14. the attenuation introduced by transmission path
15. the unit that converts sampled amplitude value to a set of pulses
16. a sequence relating to channel 1,2 and so on
17. a unique sequence of pulses called synchronization word
18. terrestrial system
19. the presence or absence of the pulse
20. a high-speed electronic switch
21. the time division multiplexer
22. time division multiplexing

三、选择合适的答案填空

1. Furthermore, we shall prove that a minimum theoretical sampling frequency of order 6.8 kHz is required ________ a voice channel ________ the range 300 Hz to 3.4 kHz.

A. convey, occupy　　　　B. to convey, occupying
C. conveying, occupied　　　　D. convey, to occupy

2. For example, the signal ________ from a satellite, ________ in far outer space, is

very weak.

A. received, located　　B. receive, locate

C. receiving, locating　　D. to receive, to locate

3. If we consider binary transmission, the complete information about a particular message will always ________ by simply ________ the presence or absence of the pulse.

A. obtain, detect　　B. be obtained, detecting

C. obtained, detected　　D. obtaining, detected

4. There is an inherent advantage for ________ noisy environments by ________ digital transmission.

A. overcoming, choose　　B. overcome, choosing

C. overcome, choose　　D. overcoming, choosing

5. Each voice channel has a separate coder, the unit ________ converts sampled amplitude values to a set of pulses; and decoder, the unit ________ performs the reverse operation.

A. who, who　　B. when, when

C. where, where　　D. that, that

6. The problem is easily overcome by ________ a frame format, where at the start of each frame a unique sequence of pulses is placed ________ the start of the frame.

A. specify, identify　　B. specifying, so as to identify

C. specified, identified　　D. specify, identifying

四、根据课文内容选择正确答案

1. If the signal is very large compared to the noise level, then ________.

A. the signal-to-noise ratio is poor

B. a perfect message can take place

C. it's a terrestrial system

D. the noise power is very strong

2. If we consider binary transmission, the complete information about a particular message will always be obtained ________.

A. by detecting the shape of the transmitted signal

B. by detecting the level of the transmitted signal

C. by calculating the parameters of the transmitted signal

D. by detecting the presence or absence of the pulse

3. At the receiving terminal ________ to separate the 8-digit sequences into the appropriate channels.

A. PCM is used

B. a demultiplexer is needed

C. the frame code is arranged

D. the coder is required

4. It is ________ which is of the most interest to the communication engineer.

A. the method for overcoming noisy environments

B. the relationship of the true signal to the noise signal

C. the binary transmission

D. the quality of the transmission

5. The codec is arranged ________, and code this value into the 8-digit sequence.

A. to do these operations

B. to perform these functions

C. to transmit the voice channels

D. to sample the amplitude value

6. Digital transmission ________ for overcoming noisy environments.

A. provides a program

B. provides a powerful method

C. provides the thermal low-level noise

D. provides the signal-to-noise ratio

7. The signal ________ is very weak.

A. within terrestrial system

B. within communication equipment

C. received from a satellite

D. in a particular message

8. The shape of the transmitted signal is most easily affected by the noise and attenuation ________.

A. introduced by the transmission path

B. introduced by the satellite

C. introduced by the equipment

D. introduced by the codec

9. The demultiplexer knows which group of 8-digits relates to channel 1 ________.

A. by choosing digital transmission

B. by identifying the synchronization word

C. by word interleaving

D. by binary transmission

五、请将下述短文译成中文

1. If we consider binary transmission, the complete information about a particular message will always be obtained by simply detecting the presence or absence of the pulse. By comparison, most other forms of transmission systems convey the message information using the shape, or level of the transmitted signal; parameters that are most easily affected by the noise and attenuation introduced by the transmission path.

Consequently there is an inherent advantage for overcoming noisy environments by choosing digital transmission.

2. The reader may ask, how does the demultiplexer know which group of 8-digits relates to channel 1, 2, and so on? Clearly this is important! The problem is easily overcome by specifying a frame format, where at the start of each frame a unique sequence of pulses called the frame code, or synchronization word, is placed so as to identify the start of the frame. A circuit of the demultiplexer is arranged to detect the synchronization word, and thereby it knows that the next group of 8-digits corresponds to channel 1.

3. Noise can be introduced into transmission path in many different ways: perhaps via a nearby lightning strike, the sparking of a car ignition system, or the thermal low-level noise within the communication equipment itself. It is the relationship of the true signal to the noise signal, known as the signal-to-noise ratio, which is of most interest to the communication engineer.

4. Basically, if the signal is very large compared to the noise level, then a perfect message can take place; however, this is not always the case. For example, the signal received from a satellite, located in far outer space, is very weak and is at a level only slightly above that of the noise. Alternative examples may be found within terrestrial systems where, although the message signal is strong, so is the noise power.

5. So far we have assumed that each voice channel has a separate coder, the unit that converts sampled amplitude values to a set of pulses; and decoder, the unit that performs the reverse operation. This need not be so, and systems are in operation where a single codec is shared between 24, 30, or even 120 separate channels.

6. A high-speed electronic switch is used to present the analog information signal of each channel, taken in turn, to the codec. The codec is then arranged to sequentially sample the amplitude value, and code this value into the 8-digit sequence. Thus the output to the codec may be seen as a sequence of 8 pulses relating to channel 1, then channel 2, and so on. This unit is called a time division multiplexer.

参考译文

PCM原理

PCM的构成依赖于三个环节,即采样、量化和编码。近年来,人们对这三个环节的实现提出了许多不同的方案,我们将对其中一些主要的方案进行讨论。在这些讨论中,我们会看到话路中的语声信号是如何转换成幅值序列的,而每个幅值又被编码,即以8位二进制数的序列表示。而且,我们将证明,为了转换频率范围为300 Hz~3.4 kHz的话路信号,理论上的最小采样频率须为6.8 kHz。但是,实际设备通常用8 kHz的采样速率,而如果每个样值用8位码的话,则话路是由一个重复速率为64 kHz的脉冲流来表示的。

图 3-1(见第 40 页)表示了采样、量化和编码的过程。

让我们再研究一下上面提到的简单例子。可以看出,最高频率为 3.4 kHz 的话音信号是用 64 kHz 的(脉冲流)信号来表示的。但是,如果每个样值只用4 位(码)表示,则传输质量将会下降,而脉冲的重复速率也将减小到 32 kHz。因而传输质量是取决于脉冲重复速率的。对于数字通信系统,这两个量之间极明显地互相影响着。

数字传输对于克服噪声环境的影响提供了一个强有力的手段。噪声可以以多种不同方式进入传输信道,比如说因为附近的闪电、汽车点火系统的打火或因通信设备本身低电平的热噪声所致。正是这种被称为信噪比的东西,即真实信号与噪声的关系引起了通信工程师的极大兴趣。本质上讲,如果信号比噪声电平大得多,则信息的传输是完美的。但是,实际情况并不总是这样,例如,从位于遥远太空中的卫星接收到的信号极其微弱,其电平仅比噪声稍高一点。地面系统则是另一类例子,尽管信号很强,噪声也很强。

通过研究二进制信号的传输可见,只要简单地去判别脉冲的"有"和"无",我们就获得了一条消息的全部信息。相比之下,许多其他形式的传输系统是利用被传信号的波形或电平的高低来传送信息的,而这些参数又极易受到传输途径中的噪声和衰耗的影响。因此选择数字传输系统在克服噪声环境的影响方面有其固有的优势。

到目前为止,在这个讨论中,我们一直假定每个话路各有一个编码器和解码器。前者是将幅度采样值变换成脉冲,而后者则施行相反的变换,这种设置并非必需。在实际运行的 PCM 系统中,一个编、译码器为 24 路、30 路,甚至 120 路所共用(注:在当代 PCM 设备中,编、译码器是分路设备,即每个话路各有一套。——本书作者)。一个高速的电子开关被用来将每一话路的模拟信号依次地送往编、译码器。然后编、译码器再顺序采样幅值并把这个幅值编成8 位码序列。这样,编解码器输出的 8 位码序列就分别对应于话路 1、话路 2 等。这种设备称为时分复用(TDM),如图 3-2 所示(见第 41 页)。由于 8 位码的码字序列按时间顺序插接在一起,所以上面所用的复用原则称为码字插接。

接收端设置了分路设备,将 8 位码序列分配到相应的话路中。读者也许会问,分路设备怎么知道哪一组 8 位码对应于第 1 路、第 2 路及其他各路呢? 显然这是很重要的。这个问题是很容易解决的。我们只要指定一个帧格式,即在每一帧的开始放置一个被称为帧码或同步字的独特的码序列以标志每帧的起始,而用分路设备的一个电路去检测同步字,从而就知道下一个 8 位码组对应于话路 1。当收到最后一个话路的码字之后,同步码字又再次出现。

UNIT 4

PASSAGE

WDM[1]

Even visionaries such as Albert Einstein and Isaac Newton, who contributed significantly to our understanding of the properties of light and its fundamental importance, would not likely imagine the communications networks of today. Highways of light span the globe, transmitting massive amounts of information in the twinkling of an eye.[2] The equivalents of millions of telephone calls are transmitted on a single fiber, thinner than a human hair. Astounding as these advances may seem, we are only at the beginning of what is possible.[3]

The current explosion of traffic in the worldwide networks is ample evidence of the speed with which we are adopting new communications technologies. The growth of wireless systems and the Internet are well-documented phenomena. No matter what application is generating traffic, most of this traffic will be carried by the unifying optical layer.[4] For this reason, the growth of various applications such as telephony (whether cellular or fixed), Internet, video transmission, computer communication and database access leads directly to an increase in the demand placed on the optical network. It is very likely that the optical network will be used to convey large amounts of video information in the future.

The most striking recent advances in optical networking have taken place in the field of wavelength division multiplexing (WDM). These advances have benefited both terrestrial and submarine systems, increased available capacities by several orders of magnitude and, correspondingly[5], reduced costs.

Until quite recently, it was possible to send only one wavelength, or color, of light along each fiber.[6] A lot of effort has therefore been concentrated in maximizing the amount of information that can be transmitted using a single wavelength. Commercial systems will soon be able to carry 40 Gbit/s on a single wavelength, while in the labs 320 Gbit/s systems have already been demonstrated.

WDM, on the other hand, makes it possible to transmit a large number of wavelengths using the same fiber[7], effectively sending a "rainbow" of colors, where there was only one color before.[8] Already today, commercially available systems can transmit 400 Gbit/s of information on a single fiber. That is equivalent to transmitting approximately 200 feature-length films per second. Recently, a team of researchers from Bell Labs demonstrated long-distance, error-free transmission of 3.28 Tbit/s over a single optical fiber.

The major advance that has led to the WDM revolution has been the invention of the optical amplifier (OA). Before the invention of the OA, wavelength had to be converted into electronic form, then back into optical form and then retransmitted into the next span of fiber. This was relatively expensive, since the optical components involved are highly specialized devices.[9] The OA, however, can boost the signal power of all wavelengths in the fiber, thus eliminating the need for separate regenerators, and allowing many wavelengths to share the same fiber. Advances in optical amplifier design have been considerable. First, the operating window has expanded from 12 nm, in the first generation, to about 80 nm today. This allows the OA to amplify more signals simultaneously. Second, the development of gain equalization techniques has enabled a more flatter response and allows a number of these amplifiers to be connected in series. There have also been advances in the fibers themselves. In the early days of optical systems, optical fibers were not built for multi-wavelength transmission. Today's fibers, however, are designed to have wide transmission windows and are optimized for high-capacity, multiple-wavelength transmission.

The growing demand on optical network is a complex issue. On the one hand, the growth in capacity demand is extraordinary, and this in itself[10] would be a big enough challenge to meet. However, this is accompanied by an increasing variety of services and applications, as well as much more exacting requirements for quality differentiation. For example, there is quite a difference in the quality requirement for a signal being used to transmit an emergency telephone call or live video coverage of a medical operation, as compared with an E-mail that is not urgent and can arrive after several hours.

However, the same optical infrastructure is expected to support this wide variety of services. Internet protocol (IP) traffic, in particular, is growing exponentially. In some parts of the world, it is expected that IP will constitute the majority of traffic in the near future. Therefore, existing networks will have to be progressively optimized to handle various types of traffic. WDM has a major advantage in this regard, which is that the different types of traffic can be assigned to different wavelengths, as required.

Fortunately, we will soon be in a position to route individual wavelengths flexibly through an optical network. Features such as add/drop and cross-connection in the optical domain are being made possible by advance in photonics.[11] I would like to draw attention to a few recent advances in this area. Firstly, the so-called digital wrapper is in the process of being standardized by the international bodies. A second significant development is the all-optical cross-connect. Bell Labs has recently unveiled its all-optical cross connect called the Lambda Router. Based on micro-electro mechanical switching (MEMS) technology, it consists of microscopic mirrors that tilt, and thus re-direct optical signals. It is such a technology that will enable us to build networks that are purely optical.[12] As more routing functions are implemented in the optical plane, more sophisticated intelligence is needed to control and manage the network. Control systems are being developed for these optical routers with which it will be able to build optical networks that can be easily configured in response to demand[13], and which also have self-healing properties and fast restoration times in the order of fifty to a hundred milliseconds, much the same as today's SDH and SONET networks.

A further aspect to consider[14] is access to the optical network. Most users would like to have direct access to the optical network and the enormous capacity it provides. This will take place in stages. Multi-wavelength optical systems are rapidly spreading out from the core towards the end user. In regional and metropolitan areas, the requirements are somewhat different from the long-distance area. The dream of fiber to the home (FTTH) or desktop is yet to materialize, mainly because of the cost-sensitive nature of this part of the network. In the near future, residential access may remain copper-based, using technologies such as ADSL to boost the capacity of traditional copper lines. However, for business offices, optical technology will be used to bring bandwidth to the end-used. Currently, a lot of fiber to the building (FTTB) networks are being deployed involving ATM and SDH access equipment at customer premises. The next step is to use WDM technology for these applications. WDM will first be used in industrial and campus local area network (LAN) environments.

We are at the beginning of a revolution in communications networks, where increasing capacity, variety of applications, and quality of service are placing enormous demands on the optical network. The revolution of optical network is just beginning, and is advancing very swiftly towards a future online world in which bandwidth is essentially unlimited, reliable and low-cost.

NEW WORDS AND PHRASES

visionary	*n*. 幻想家,梦想家
contribute	*v*. 捐献,贡献
significant	*a*. 重大的,效果显著的,具有特殊意义的
property	*n*. 所有物,房产,所有权
fundamental	*a*. 基础的,重要的,本性的
span	*v*. 跨过,延伸
twinkle	*vi*. 闪烁,闪耀,眨眼
equivalent	*a*. 相同的,同等
traffic	*n*. 交通,通信量,交易
ample	*a*. 充足的,充分的,宽敞的
evidence	*n*. 证据,证词

phenomena	*n*. 现象(单数为 phenomenon)
unify	*v*. 使一致,使化一,统一
video	*a*. 电视的,视频的,录像的
striking	*a*. 引人注意的,显著的
advance	*vi*. 前进,提升,进展
magnitude	*n*. 广大,巨大,重要
corresponding	*a*. 符合的,对应的,一致的
commercial	*a*. 商业的,商务的
demonstrate	*v*. 示范,展示,演示
rainbow	*n*. 虹,彩虹
approximate	*a*. 近似的,大概的
boost	*v*. 上推,增加,提高
eliminate	*vt*. 除去,淘汰
considerable	*a*. 相当大的,相当多的
simultaneous	*a*. 同时发生的,同时存在的,同时的
equalization	*n*. 相等,均等,平等
optimize	*v*. 乐观的考虑,使尽可能完善
extraordinary	*a*. 非常的,特别的,非凡的
exacting	*a*. 苛求的,严格的
differentiation	*n*. 分化,变异,演变
medical	*a*. 医学的,医术的,医疗的
infrastructure	*n*. 基础,基础结构
exponential	*a*. 指数的,幂的
constitute	*vt*. 构成,组成,设立
majority	*n*. 多数,半数以上
in this regard	在这一点上,关于此事
flexible	*a*. 易弯曲的,可变通的,灵活的
domain	*n*. 领域,领地,范围
wrapper	*n*. 包装者,包装物,覆盖物
standardize	*vt*. 使与标准比较,使合标准,使标准化
twofold	*a*. 两倍的,两重的
identify	*v*. 使等同于,识别,鉴定
unveil	*v*. 使公之于众,揭露,展出
microscopic	*a*. 显微镜的,微小的,细微的
implement	*vt*. 贯彻,完成,履行

sophisticate	v. 使复杂,使精致
intelligence	n. 智力,智能,理解力
router	n. 路由器
restoration	n. 恢复,复位,复原
aspect	n. 样子,外表,方面
enormous	a. 巨大的,庞大的
metropolitan	a. 大城市的,大都会的
materialize	v. 使物质化,使具体化
residential	a. 居住的,长住的,居留的
deploy	v. 展开,调度,部署

NOTES

1. 本篇课文涉及光纤通信技术,题目可译成:波分复用(WDM)。

2. transmitting massive… 分词短语在此表示结果状语,表示一种伴随的情况。可翻译为"在眨眼间传输巨量的信息"。

3. astounding as these…,as 在此引导让步状语从句,可译成"尽管、虽然","尽管这些发展使人惊讶"。as 的用法比较复杂,在后面的阅读文章中还会经常遇到。

4. no matter what… 引导让步状语从句,译成"无论什么"。

5. correspondingly 插入语,译成"因而、相应地"。

6. it 做形式主语时,可以代替不定式短语,真正的主语是后面的不定式短语。这时 it 只是形式上的主语,本身没有词汇意义,不必译出,句型为"It is (was)+形容词+不定式"。

7. makes it possible to transmit…,it 在此是形式宾语,没有词汇意义,代替后面的不定式短语 to transmit。当动词不定式短语或从句在句中做宾语,而这种宾语又带有补足语时,通常要用 it 作为引导词放在宾语补足语之前,而把真正的宾语,即不定式短语或从句放在补足语之后。

8. where there was only…, where 在此引导地点状语从句,译为"而在以前只有一种颜色"。

9. since 引导原因状语从句,译为"因为……"。since 只表示事物内在联系上的一种合乎逻辑的自然结果。从语气上来说,because 语气最强,since 次之,for 最弱。

10. in itself 译为"本身"。

11. are being made 为被动语态,其构成为"助动词 be+过去分词",时态通过 be 的各种形式来体现,在英语中共有 10 种形式。在此为被动语态进行时。

12. It is… 强调句,句型为"It is (was)+被强调的成分+that(which, who)",一般可用"正是……"来表示。

13. with which… 句型为"介词+which+句子其他成分"。一般有几种形式,在此构成状语,that 引出定语从句,并在从句中做主语。

14. 不定式做后置定语，修饰 aspect。

EXERCISES

一、请将下述词组译成英文

1. 对光特性的理解
2. 基本重要性
3. 想象今天的通信系统
4. 光的高速公路
5. 巨量的信息
6. 采用通信新技术
7. 大量的视频信息
8. 波分复用
9. 只发送单个波长
10. 传输大量的波长
11. 无差错传输
12. 自愈特性
13. 直接接入光网络
14. 视频信息

二、请将下述词组译成中文

1. the major advance that led to the WDM revolution
2. the invention of the optical amplifier
3. the next span of fiber
4. to boost the signal power of all wavelength
5. the advances in optical amplifier
6. the development of gain equalization techniques
7. the multiple-wavelength transmission
8. the growth of wireless system
9. the growth of various application
10. the wide variety of services
11. to handle various types of traffic
12. the all-optical cross-connect

三、选择合适的答案填空

1. Highways of light span the globe, ________ massive amounts of information in the twinkling of an eye.

A. to transmit　　B. transmitting
C. transmitted　　D. be transmitted

2. The equivalents of millions of telephone calls are ________ on a single fiber, thinner than a human hair.

A. transmitted　　B. transmitting
C. to transmitted　　D. to transmit

3. WDM, on the other hand, makes it possible ________ a large number of wavelengths using the fiber, effectively sending a "rainbow" of color, where there was only one color before.

A. to transmit　　B. to transmitting
C. transmitted　　D. to transmitted

4. Before the invention of the OA, after having traveled down a fiber for some distance, each individual wavelength had to be ________ into electronic form, then back

into optical form and then retransmitted into the next span of fiber.

A. to convert B. converted

C. converting D. to converting

5. In the near future, residential access may remain copper-based, ________ technologies such as ADSL to boost the capacity of traditional copper lines.

A. to use B. use

C. using D. used

6. However, the same optical infrastructure is expected ________ this variety of services.

A. to support B. supporting

C. supported D. to be support

7. It is very likely that the optical network will be ________ to convey large amounts of video information in the future.

A. to use B. to used

C. using D. used

四、根据课文内容选择正确答案

1. As more routing functions are implemented in the optical plane, we need ________.

A. more sophisticated intelligence to control and manage the network

B. self-healing properties and fast restoration times

C. direct access to the optical network

D. live video coverage of a medical operation

2. Therefore, to handle various types of traffic, ________.

A. this in itself would be big enough challenge to meet

B. this allow the OA to amplify more signals simultaneously

C. today's network will gave to be progressively optimized

D. the different types of traffic can be assigned to different wavelength, as required

3. Second, the development of gain equalization techniques has enabled a much flatter response and allow ________.

A. much more exacting requirements for quality differentiation

B. a number of these amplifiers to be connected in series

C. the majority of traffic in the near future

D. our understanding of the properties of light

4. For this reason, the growth of various applications, such as telephony, Internet, video transmission, computer communication and database access________ on the network.

A. would not likely to imagine the communication network of today

B. be only at the beginning of what is possible

C. have already been demonstrated

D. leads directly to an increase in the demand placed

5. The major advance that has led to the WDM revolution ________.

A. can boost the power of all wavelengths

B. will be used to convey large amount of video information in the future

C. has been the invention of OA

D. is access to the optical network

6. As more routing functions are implemented in optical plane, more sophisticated intelligence ________.

A. are being developed for these optical router

B. can be easily configured in response to demand

C. are being made possible by advance in photonics

D. is needed to control and manage the network

五、请将下述短文译成中文

1. The current explosion of traffic in the worldwide networks is ample evidence of the speed with which we are adopting new communications technologies. The growth of wireless systems and the Internet are well-documented phenomena.

2. The most striking recent advances in optical networking have taken place in the field of wavelength division multiplexing. These advances have benefited both terrestrial and submarine systems, increased available capacities by several orders of magnitude and, correspondingly, reduced cost.

3. The OA, however, can boost the signal power of all wavelengths in the fiber, thus eliminating the need for separate regenerators, and allowing more wavelengths to share the same fiber.

4. For example, there is quite a difference in the quality requirement for a signal being used to transmit an emergency telephone call or live video cover-age of a medical operation, as compared with an E-mail that is not urgent and can arrive after hours.

5. A further aspect to consider is access to the optical network. Most users would like to have direct access to the optical network and the enormous capacity it provides. This will take place in stages. Multi-wavelength optical systems are rapidly spreading out from the core towards the end user.

6. Wavelength division multiplexing (WDM) is an optical technology that couples many wavelengths in the same fiber, thus effectively increasing the aggregate bandwidth per fiber to the sum of the bit rates of each wavelength. As an example, 40 wavelengths at 10 Gbit/s per wavelength in the same fiber raise the aggregate bandwidth to 400 Gbit/s, and astonishing aggregate bandwidths at several terabits per second (Tbit/s) are also a reality. As opto-electronic technology moves forward, it is possible to have a high density of wavelengths in the same fiber. Thus, the term dense wavelength division

multiplexing (DWDM) is used. Dense WDM (DWDM) is a technology with a larger (denser) number of wavelengths (>40) coupled into a fiber than WDM.

7. Conventional single-mode fibers transmit wavelengths in the 1 310 nm and 1 550 nm ranges and absorb wavelength in the range 1 340 nm to 1 440 nm range. WDM systems use wavelengths in the two regions of 1310 nm and 1 550 nm. Special fibers have made it possible to use the complete spectrum from 1310 to beyond 1600 nm. However, although new fiber technology opens up the spectrum window, not all optical components perform with the same efficiency over the complete spectrum. As an example, erbium-doped fiber amplifiers perform best in the range of 1 550 nm.

8. Currently, commercial systems with 16, 40, 80 and 128 channels (wavelengths) per fiber have been announced. Those with forty channels have channel-spacing of 100 GHz, and those with 80 channels have channel spacing at 50 GHz. Forty-channel DWDM systems can transmit an aggregate bandwidth of 400 Gbit/s over a single fiber. It is estimated that at 400 Gbit/s, more than 10 000 volumes of an encyclopedia can be transmitted in 1 second.

9. Although DMDM technology is still evolving and technologists and standards bodies are addressing many issues, systems are being offered have only a few dozen wavelengths. But there are reasons to believe that in the near future we will see DWDM systems with several hundreds of wavelengths in a single fiber. Theoretically, more than 1 000 channels may be multiplexed in a fiber. DWDM technology with more than 200 wavelengths has already been demonstrated.

参考译文

波分复用(WDM)

爱因斯坦和牛顿在我们对光特性的了解及其基本重要性方面作出了重大贡献,即使是这样富于幻想的人也不大可能想象得到今天通信网络发展的程度。光的高速公路跨越全球,在眨眼之间传输巨量的信息。这等于数百万电话的通信量通过比人头发还细的光纤进行传输。这些成就似乎使人很吃惊,但我们只是处于可能会产生更大成就的开始。

目前全球网络业务爆炸式增长充分地体现了我们采用新的通信技术的速度。无线系统和因特网的发展就是很好的证明。不管是哪一种应用在产生流量,大多数这种业务将会由统一的光层进行传送。由于这一原因,各种应用的增长,如电话(移动或固定)、因特网、视频传输、计算机通信和数据库接入,直接导致对光网络持续增长的需要。极有可能,将来光网络会用来传送大量的视频信息。

近来在光网络方面最引人注目的进展出现在波分复用(WDM)方面。这些进展给地面通信系统和海底通信系统带来了好处,将可用的容量增加了几个数量级,而相应地减少了费用。

目前,在单根光纤上只能发送单波长或单颜色的光。因此,大量的努力集中在使用单个波长传输信息数量的最大化方面。商用系统很快能够在单个光纤上传输 40 Gbit/s 的信息，而在实验室，已经演示了传输 320 Gbit/s 的系统。

另一方面，WDM 使得使用同样的光纤传输很多的波长成为可能，可能有效地发送颜色的“彩虹”，而在以前只有一种颜色。今天商用系统可以在一根光纤上传输 400 Gbit/s 的信息，这大约等效于每秒传输 200 部故事片。最近，贝尔实验室的一个研究队伍展示了一根光纤速率高达 3.28 Tbit/s 的长途、无误码的传输系统。

导致 WDM 革命的主要进展是光放大器（OA）的发明。在光放大器发明之前，单个的波长在经过光纤传输一定距离之后，必须转换成电信号的形式，然后再变换回到光信号形式并通过下一段光纤传输。这样就使得费用比较昂贵，因为所涉及的光器件是高度专用的器件。然而，光放大器可以放大光纤中所有波长的信号功率，这样就不需要不同的再生器，并允许许多波长共用同样的光纤。在光放大器设计方面的进展具有不可忽视的重要作用。首先，光纤的工作窗口已由第一代的 12 nm 扩展到今天的 80 nm。这使光放大器可以同时放大更多的信号。第二，增益均衡技术的发展可以产生更平坦的响应，并允许这样的一些放大器以串联的形式连接。光纤本身也有了进展。在早期的光系统中，敷设的光纤不能用于多波长传输。然而，今天设计的光纤有宽的传输窗口，可以优化地进行高容量、多波长传输。

对光网络日益增长的需求是一个复杂的问题。一方面对容量需求的增长是异乎寻常的，这本身就是一个需要面对的很大挑战。然而同时伴随的是日益增长的各种业务和应用以及对不同质量业务更加严格的需要。例如，用于传输紧急电话或者实况医疗手术转播的信号与用于发送不是很紧急的可以几个小时后到达的电子邮件的信号相比，信号质量有很大的不同。

然而，人们同样希望有光网络结构来支持这种广泛的业务。特别是 IP 业务，正在以指数形式增长。在世界的一些地方，IP 业务有可能在不远的将来构成业务中的大部分。因此，现有的网络必须逐步进行优化以处理各类型的业务。WDM 在这一方面有很大的优势，这种优势就是不同类型的业务按照需要可以分配不同的波长信号。

幸运的是，我们很快就可以通过光网络灵活地传送单个波长的信号。光子学领域的发展使得在光域上具有的信号上下和交叉连接的特性正在成为可能。我想让大家注意这一领域的几个最新的进展。首先，国际标准化组织正在对所谓的数字封包技术进行标准化工作。第二个重大的进展是全光交叉连接。贝尔实验室最近公布了其称作 Lambda Router 的全光交叉连接研究成果。Lambda Router 技术是以微电子机械交换（MEMS）技术为基础，由倾斜的极微小的镜子组成，这样就可以重新引导光信号。正是这种技术将会使我们能够建立一个全光网络。随着越来越多的路由选择功能在光平台实现，就需要更复杂的智能来控制和管理这种网络。用于光路由器的控制系统正在开发中，用这种路由器可以很容易地根据需要来配置光网络，它还具有自愈特性和数量级为50～100 ms的快速恢复时间，这与今天的 SDH 和 SONET 网络非常一致。

进一步要考虑的是接入光网络。多数用户想要直接接入光网络以利用光网络所提供的巨大的信息容量。这将会分成几个阶段来完成。多波长光系统正快速地从核心网向终端用户扩展。在大区和城区的需求与长途通信有些不同。光纤到家（FTTH）或光纤到桌面的梦想还需要经过努力去实现,主要是由于这一部分网络的费用敏感性。在近来一些日子里,住户接入可能还是采用铜线,可以使用ADSL技术以增加传统铜线的容量。然而,对于商用办公室,可采用光纤技术将带宽送到终端用户。目前,很多光纤到建筑物(FTTB)网络正在敷设中,这包括在用户房屋中的ATM和SDH接入设备。下一步是在这些应用中使用WDM技术。WDM将会首先被使用在工业和校园局域网环境中。

我们正处于通信网络革命的开始,越来越大的容量需求,各种各样的应用以及服务质量正在对光网络提出巨大的需求。光纤的革命刚刚开始,并正快速地向未来带宽无限的、可靠的、低费用的在线世界发展。

UNIT 5

PASSAGE

5G[1] and Its Application

5G (from "5th Generation") is the latest generation of cellular mobile communications. 5G performance targets high data rate, reduced latency, energy saving, cost reduction, higher system capacity, and massive device connectivity. The first phase of 5G specifications in Release-15 will be completed by April 2019 to accommodate the early commercial deployment.[2] The second phase in Release-16 is due to be completed by April 2020 for submission to the International Telecommunication Union (ITU) as a candidate of IMT-2020 technology. 5G will connect everything, and benefit all walks of life. It will combine big data, cloud computing, artificial intelligence, and many other innovative technologies to accelerate the arrival of a golden age of information over the next 10 years.

With the tide of IoT(Internet of Things), more and more mobile smart devices besides smart phones will access the mobile Internet. The next generation mobile communication system, which is also known as the 5TH generation (5G), will offer three types of scenarios[3]: first, the enhanced mobile broadband (eMBB) aims to provide broadband multimedia to human-centric use cases; second, the ultra-reliable low latency service (URLLC) with strict requirements in terms of latency (ms level) and reliability (five nines and beyond) is used for remote control of robots or tactile Internet applications; third, massive machine type communications (mMTC), is mainly used to connect a very large number of devices and transmit a low volume of non-delay sensitive information. It is believed that the total throughput will grow 1 000 times from 2010 to 2020 and the number of devices will grow to 500 billion.[4]

In order to achieve the capacity growth, 5G cells have to be densely deployed, about 40 to 50 times as many as 4G networks. Because of the application of new air interface techniques, varied services and terminals, a typical 5G node has about 2 000 parameters to be configured. 5G planning also aims at lower latency and lower energy consumption, for better implementation of Internet of Things (IoT). More specifically, there are eight advanced features of 5G wireless systems: 1 – 10 Gbit/s connections to end points in the field, 1 millisecond latency, 1 000 bandwidth per unit area, 10 – 100 number of connected devices, 99.999% availability, 100% coverage, 90% reduction of network energy usage and up to ten-years battery life for low power devices.

5G networks achieve these higher data rates by using higher frequency radio waves, in the millimeter wave band around 28 GHz and 39 GHz while previous cellular networks used frequencies in the microwave band between 700 MHz and 3 GHz.[5] A second lower frequency range in the microwave band, below 6 GHz, will be used by some providers, but this will not have the high speeds of the new frequencies. Because of the more plentiful bandwidth at these frequencies, 5G networks will use wider frequency channels to communicate with the wireless device, up to 400 MHz compared with 20 MHz in 4G LTE, which can transmit more data (bits) per second.[6] OFDM (orthogonal frequency division multiplexing) modulation is used, in which multiple carrier waves are transmitted in the frequency channel, so multiple bits of information are being transferred simultaneously, in parallel.

Millimeter waves are absorbed by gases in the atmosphere and have shorter range than microwaves, and therefore the cells are limited to smaller size; 5G cells will be the size of a city block, as opposed to the cells in previous cellular networks which could be many miles across.[7] The waves also have trouble passing through building walls, requiring multiple antennas to cover a cell. Millimeter wave antennas are smaller than the large antennas used in previous cellular networks, only a few inches long, so instead of a cell tower 5G cells will be covered by many antennas mounted on telephone poles and buildings.[8]

Another technique used for increasing the data rate is massive MIMO (multiple-input multiple-output).[9] Each cell will have multiple antennas communicating with the wireless device, each over a separate frequency channel, received by multiple antennas in the device, and thus multiple bitstreams of data will be transmitted simultaneously, in parallel. In a technique called beamforming the base station computer will continuously calculate the best route for radio waves to reach each wireless device, and will organize multiple antennas to work together as phased arrays to create beams of millimeter waves to reach the device.[10] The smaller, more numerous cells makes 5G network infrastructure more expensive to build per square kilometer of coverage than previous cellular networks. Deployment is currently limited to cities, where there will be enough users per cell to provide an adequate investment return, and there are doubts about whether this technology will ever reach rural areas.

5G—in combination with other technologies such as artificial intelligence (AI), Internet of Things (IoT), augmented reality (AR), virtual reality (VR), and blockchain—is likely to be a transformative force in the e-commerce industry and market.

Internet of Things

A study found that 70 percent of retailers worldwide were ready to adopt IoT to improve consumer experience.[11] There are a number of ways in which e-commerce activities can benefit from IoT. For instance, IoT makes it easy to track inventory in real time and manage it more effectively. By doing so, human errors can be reduced. IoT can also help minimize waste, control costs, and reduce shortage. Vast amounts of unstructured data are created by IoT devices, and the amount of data created is growing twice as fast as the available bandwidth. It is estimated that by 2020, a network capacity that is at least 1 000 times the level of 2016 will be needed.[12] The amount of communication that needs to be handled will also increase exponentially.

The Internet of Things is the foundation for the commercial use of 5G and the purpose of developing 5G is to bring convenience to our production and life. And the Internet of things provides a platform for 5G to fulfill its potential. On this stage, 5G can play a powerful role through many IoT applications, such as smart agriculture, smart logistics, smart home, Internet of vehicles, smart city, etc.

Virtual and Augmented Reality

VR and AR are likely to emerge as driving forces in the e-commerce industry and market. By wearing a VR headset, a shopper can instantly find herself in a company's virtual shop, where she can "walk" around to explore items exactly as she would in the real shop.[13] For instance, if she wants to know more about a new piece of jewelry in the shop, she can focus her sight on that item and see the relevant information needed to make a purchasing decision. If she wants to buy it, she can make the payment or add it to her cart and look for additional items.

The complexity and richness of the AR and VR worlds require processing a large quantity of data.[14] Current 4G networks standards suffer from some limitations such as those related to bandwidth, latency, and uniformity, especially when the data needs to be fed remotely. In this regard, 5G is likely to unlock the full potential of VR and AR technologies. 5G's significantly faster speeds and lower latency would help overcome these weaknesses. 5G streams' transmission delay is about 1 millisecond, which is much shorter than human beings can notice.[15] 5G's almost zero delay in transmission is thus likely to enrich customers' experience with AR and VR technologies.

Blockchain

Sophisticated applications of blockchain to facilitate e-commerce activities have been or are being developed. JD.com has implemented blockchain in its supply-chain management system and B2B e-commerce. In 2017, the system went live with beef manufacturer Kerchin as its first supply-chain partner. The company announced that JD.com would have more than 10 brands of alcohol, food, tea, and pharmaceutical products on its blockchain by the end of October 2017.[16]

One of the most high-profile future uses of blockchain is likely to be smart contracts. Online vendors can use smart contracts to automate fulfillment of orders for the delivery of digital products. In smart contracts that are executed on the blockchain, 5G can play a key role in feeding the information (for example, from IoT devices) more efficiently.[17]

AI

AI-enabled devices are already playing important roles in helping consumers in e-commerce activities such as making buying decisions and tracking products. For instance, virtual assistants are transforming the way consumers purchase products online. Amazon's personal assistant Alexa has been integrated into Amazon products as well as those from other manufacturers. Customers can use Alexa to find information about local concerts through eBay's online ticket exchange company StubHub. In addition, they can arrange transportation to the event via Uber and order pre-event dinner from Domino's. The order status can be tracked in real time.

5G will dramatically improve consumer experience with AI-based devices. With 5G, AI-based devices can access additional structured and unstructured information and better understand the environment. Overall, AI-powered services will be more reliable in a wide range of contexts and situations in which they operate.

As a leader in 5G technologies, Huawei has completed interoperability development testing (IODT) with mainstream chip, terminal, and network vendors. Huawei became the first company worldwide to launch the industry-first 5G commercial chip with the Balong 5G01 and 5G commercial CPE compliant with 3GPP Release 15. Huawei is the only vendor who can provide end-to-end commercial solutions, vigorously promoting the maturity and commercial use of the 5G industry chain.[18]

Some of the current challenges in the development of e-commerce can be overcome with the deployment of 5G networks, such as with IoT devices. Addressing the exponential growth in IoT devices will be no small feat. Current 4G networks, however, cannot handle all the data coming from IoT devices. 5G's higher data transmission and processing speeds will address this concern. Specifically, 5G (in combination with AI, VR, AR, and other technologies) will play a powerful role in transforming the e-commerce industry and market. Such a combination can result in a rich e-commerce ecosystem and a better customer experience.

NEW WORDS AND PHRASES

performance	*n*. 性能,绩效,表演
latency	*n*. 延迟,时延,潜伏
massive	*a*. 大量的,大规模的,巨大的
connectivity	*n*. 连通性,互联互通
specification	*n*. 规格,说明书,详述
submission	*n*. 屈服,提交,服从
benefit	*v*. 对……有利益,对……有好处
artificial	*a*. 人工的,人造的

intelligence	*n*. 智力，智能，情报
innovative	*a*. 创新的革新的
smart	*a*. 聪明的，机灵的，巧妙的
scenario	*n*. 场景，情节，脚本
robot	*n*. 机器人，机械手，遥控设备
tactile	*a*. 触觉的，有触觉的
sensitive	*a*. 敏感的，感觉的，易受影响的
interface	*n*. 界面，接口
parameter	*n*. 参数，系数，参量
configure	*vt*. 配置，设置
consumption	*n*. 消费，消耗
implementation	*n*. 实现，实施，执行
battery	*n*. 电池，蓄电池
plentiful	*a*. 丰富的，许多的
orthogonal	*a*. 正交的，直角的
modulation	*n*. 调制，调整
simultaneously	*ad*. 同时地
absorb	*vt*. 吸收，吸引
atmosphere	*n*. 气氛，大气
antenna	*n*. 天线，触角
beamform	*v*. 波束形成
phase	*n*. 相位，阶段
array	*n*. 数组，阵列，排列
infrastructure	*n*. 基础设施，基础结构
blockchain	*n*. 区块链
augment	*v*. 增加，增大
virtual	*a*. 虚拟的
estimate	*v*. 估计，估量
exponentially	*ad*. 指数地，按指数方式地
cart	*n*. 小车，购物车
complexity	*n*. 复杂，复杂性
uniformity	*n*. 均匀性，一致性
sophisticated	*a*. 复杂的，精细的，精致的
facilitate	*vt*. 促进，使容易
context	*n*. 环境，上下文，语境

interoperability	*n*. 互操作性,互用性
chip	*n*. 芯片,晶片
maturity	*n*. 成熟期,成熟

NOTES

1. 5G代表第五代移动通信。

2. 本句为被动句,可以按主动句来翻译。to accommodate the early commercial deployment 引导不定式短语,表示目的。译文:Release-15会在2019年4月完成5G规范的第一阶段,以适应早期的商业部署。

3. which is also known as the 5th generation (5G)引导非限定性定语从句;be known as,被称作。译文:下一代移动通信系统,也称作第五代系统,将会提供三种场景类型。

4. it是形式主语;由and连接两个并列句;times表示倍数。译文:相信从2010年到2020年总吞吐量将会增长1 000倍,连接设备数量增长到5 000亿。

5. by using higher frequency radio waves表示方式;while做连词(对比两事物,……而)。译文:5G网络通过使用更高的无线频率来实现这些更高的数据速率,毫米波频段在28 GHz至39 GHz之间,而之前的蜂窝网络使用的频率范围在700 MHz至3 GHz之间的微波频段。

6. because of,由于;which can transmit more data (bits) per second为非限定性定语从句。译文:由于这些频率的带宽更丰富,5G网络将会使用更宽的频率通道与无线设备通信,最高可达400 MHz,而4G LTE为20 MHz,每秒可传输更多的数据比特。

7. as opposed to 与……截然相反;which could be many miles across为限定性定语从句,可采用分译法翻译。译文:5G的小区范围有街区那么大,而以前的蜂窝网络可能有好几英里宽。

8. smaller than为比较级,小于;used in previous cellular networks过去分词短语做后置定语,修饰前面的antenna;mounted on telephone poles and buildings过去分词短语做后置定语,修饰前面的antenna。译文:毫米波天线比以前的蜂窝网络中使用的大型天线要小,只有几英寸长,所以5G小区将被安装在电线杆和建筑物上的许多天线覆盖,而不是发射塔。

9. used for increasing the data rate 过去分词短语做后置定语,修饰前面的technique;MIMO,多输入多输出。译文:另一种提高数据速率的技术是大规模MIMO(多输入多输出)。

10. called beamforming过去分词短语做后置定语,修饰technique;and连接两个并列句;as,作为;to create beams of millimeter waves不定式短语表示目的。译文:在一种称为波束形成的技术中,基站计算机将不断计算无线电波到达每个无线设备的最佳路径,并将组织多个天线作为相控阵列一起工作,产生毫米波波束到达设备。

11. that引导宾语从句,作found的宾语,表示found的内容;to improve consumer

experience 不定式短语表示目的。译文:一项研究发现,全球 70%的零售商已准备好采用物联网来改善消费者体验。

12. it 是形式主语;that is at least 1,000 times the level of 2016 是定语从句,修饰 capacity。译文:据估计,到 2020 年,将需要一个至少是 2016 年水平 1 000 倍的网络容量。

13. by wearing a VR headset 介词短语,表示方式;where 引导定语从句,修饰前面的 shop: as she would in the real shop,as 引导时间状语从句。译文:通过佩戴 VR 装置,购物者可以立即发现自己在公司的虚拟商店中,在那里她可以"走动"来搜索与在真实商店中完全一样的物品。

14. The complexity and richness of the AR and VR worlds 名词短语做句子主语。译文:AR 和 VR 世界的复杂性和丰富性需要处理大量数据。

15. which is much shorter than human beings can notice,非限定性定语从句;shorter than 为比较级,much 表示强调。译文:5G 数据流的传输延迟大约是 1 毫秒,这比人类可以注意到的时长要短得多。

16. that JD. com would have more than 10 brands of alcohol, food, tea, and pharmaceutical products,that 引导宾语从句,做动词 announced 的宾语。译文:该公司宣布,京东将于 2017 年 10 月底在其区块链上拥有超过 10 个品牌的酒精饮料、食品、茶叶和药品。

17. that are executed on the blockchain 定语从句,修饰 contract;play a key role,发挥重要作用。译文:在区块链上执行的智能合约中,5G 可以在更有效地提供信息(例如,来自物联网设备)方面发挥关键作用。

18. who 引导定语从句,修饰 vendor; vigorously promoting the maturity and commercial use of the 5G industry chain 分词短语,修饰整个句子,前面有副词 vigorously 修饰。译文:华为也是能够提供端到端商用解决方案的唯一的销售商,积极的促进 5G 产业链的成熟和商业应用。

EXERCISES

一、请将下述词组译成英文

1. 更高的系统容量	2. 大规模设备连接	3. 连接万物
4. 许多其他创新技术	5. 移动智能设备	6. 接入移动互联网
7. 增强的移动宽带	8. 机器人远程控制	9. 大规模机器通信
10. 多输入多输出(MIMO)	11. 覆盖不同的频带	12. 称作波束形成的技术
13. 戴上 VR 头盔	14. 实现容量增长	15. 密集部署
16. 更低的能耗	17. 在毫米波频带	18. 早先的蜂窝网络
19. 毫米波天线	20. 公司的虚拟商店	21. 改进消费者体验

二、请将下述词组译成中文

1. the ultra-reliable low latency service
2. the first phase of 5G specifications
3. a candidate of IMT-2020 technology

4. to accelerate the arrival of a golden age of information
5. to accommodate the early commercial deployment
6. requiring multiple antennas to cover a cell
7. the large antennas used in previous cellular networks
8. many antennas mounted on telephone poles and buildings
9. the application of new air interface techniques
10. to communicate with the wireless device
11. another technique used for increasing the data rate
12. to organize multiple antennas to work together
13. to provide an adequate investment return
14. the available bandwidth
15. vast amounts of unstructured data
16. the amount of data created
17. the foundation for the commercial use of 5G
18. the complexity and richness of the AR and VR worlds
19. the full potential of VR and AR technologies
20. almost zero delay in transmission
21. to improve consumer experiences with AI-based devices
22. to launch the industry-first 5G commercial chip
23. to result in a rich e-commerce ecosystem

三、选择合适的答案填空

1. 5G performance targets high data rate, ________ latency, energy saving, cost reduction, higher system capacity, and massive device connectivity.

A. to reduce　　B. reduced
C. reduce　　D. reducing

2. The second phase in Release-16 is due to be ________ by April 2020 for submission to the International Telecommunication Union (ITU) as a candidate of IMT-2020 technology.

A. complete　　B. completing
C. to complete　　D. completed

3. Third, massive machine type communications (mMTC), mainly ________ to connect a very large number of devices and transmitting a low volume of non-delay-sensitive information.

A. used　　B. using
C. to use　　D. to be used

4. A second lower frequency range in the microwave band, below 6 GHz, will be ________ by some providers, but this will not have the high speeds of the new frequencies.

A. to use　　B. use

C. used　　D. using

5. The waves also have trouble passing through building walls, ________ multiple antennas to cover a cell.

A. to require　　B. to be required

C. require　　D. requiring

6. Each cell will have multiple antennas ________ with the wireless device, each over a separate frequency channel, received by multiple antennas in the device, thus multiple bitstreams of data will be transmitted simultaneously, in parallel.

A. to communicate　　B. communicate

C. communicated　　D. communicating

7. For instance, IoT makes it easy ________ inventory in real time and manage it more effectively.

A. to track　　B. track

C. tracked　　D. tracking

8. Vast amounts of unstructured data are ________ by IoT devices, and the amount of data created is growing twice as fast as the available bandwidth.

A. to be cteated　　B. creat

C. created　　D. creating

四、根据课文内容选择答案

1. The second phase in Release-16 is due to be completed by April 2020 for submission to the International Telecommunication Union (ITU) as ________.

A. a golden age of information

B. a candidate of IMT-2020 technology

C. new air interface techniques

D. better implementation of Internet of Things

2. 5G will connect everything, and benefit all walks of life. It will combine big data, cloud computing, artificial intelligence, and ________ to accelerate the arrival of a golden age of information over the next 10 years.

A. many other innovative technologies

B. massive device connectivity

C. the first phase of 5G specifications

D. more mobile smart devices

3. In order to ________, 5G cells have to be densely deployed, about 40 to 50 times as many as 4G networks.

A. communicate with the wireless device

B. end points in the field

C. provide broadband multimedia

D. achieve the capacity grows

4. OFDM (orthogonal frequency division multiplexing) modulation is used, in which multiple carrier waves are ________, so multiple bits of information are being transferred simultaneously, in parallel.

A. used for remote control of robots

B. absorbed by gases in the atmosphere

C. transmitted in the frequency channel

D. completed by April 2019

5. Because of the ________ techniques, varied services and terminals, a typical 5G node has about 2 000 parameters to be configured.

A. first phase of 5G specifications

B. more and more mobile smart devices

C. the ultra-reliable low latency service

D. application of new air interface

6. 5G planning also aims at lower latency and ________, for better implementation of Internet of Things (IoT).

A. lower energy consumption

B. new air interface techniques

C. low volume of non-delay-sensitive information

D. massive device connectivity

7. The waves also have trouble passing through building walls, requiring multiple antennas ________.

A. to build per square kilometer of coverage

B. to cover a cell

C. to create beams of millimeter waves

D. to track inventory in real time

五、请将下列短文译成中文

1. The second phase in Release-16 is due to be completed by April 2020 for submission to the International Telecommunication Union (ITU) as a candidate of IMT-2020 technology. 5G will connect everything, and benefit all walks of life. It will combine big data, cloud computing, artificial intelligence, and many other innovative technologies to accelerate the arrival of a golden age of information over the next 10 years.

2. With the tide of IoT(Internet of Things), more and more mobile smart devices besides smart phones will access to the mobile Internet. The next generation mobile communication system, which is also known as the 5TH generation (5G), will offer three types of scenarios: first, the enhanced mobile broadband (eMBB) aims to provide broadband multimedia to human-centric use cases.

3. In order to achieve the capacity grows, 5G cells have to be densely deployed, about 40 to 50 times as many as 4G networks. Because of the application of new air interface techniques, varied services and terminals, a typical 5G node has about 2 000 parameters to be configured. 5G planning also aims at lower latency and lower energy consumption, for better implementation of Internet of Things (IoT).

4. 5G networks achieve these higher data rates by using higher frequency radio waves, in the millimeter wave band around 28 GHz and 39 GHz while previous cellular networks used frequencies in the microwave band between 700 MHz and 3 GHz. A second lower frequency range in the microwave band, below 6 GHz, will be used by some providers, but this will not have the high speeds of the new frequencies. Because of the more plentiful bandwidth at these frequencies, 5G networks will use wider frequency channels to communicate with the wireless device, up to 400 MHz compared with 20 MHz in 4G LTE, which can transmit more data (bits) per second.

5. Millimeter waves are absorbed by gases in the atmosphere and have shorter range than microwaves, and therefore the cells are limited to smaller size; 5G cells will be the size of a city block, as opposed to the cells in previous cellular networks which could be many miles across. The waves also have trouble passing through building walls, requiring multiple antennas to cover a cell.

6. Another technique used for increasing the data rate is massive MIMO (multiple-input multiple-output). Each cell will have multiple antennas communicating with the wireless device, each over a separate frequency channel, received by multiple antennas in the device, and thus multiple bitstreams of data will be transmitted simultaneously, in parallel. In a technique called beamforming the base station computer will continuously calculate the best route for radio waves to reach each wireless device, and will organize multiple antennas to work together as phased arrays to create beams of millimeter waves to reach the device.

7. The Internet of things is the foundation for the commercial use of 5G and the purpose of developing 5G is to bring convenience to our production and life. And the Internet of Things provides a platform for 5G to fulfill its potential. On this stage, 5G can play a powerful role through many IoT applications, such as smart agriculture, smart logistics, smart home, Internet of vehicles, smart city, etc.

8. The complexity and richness of the AR and VR worlds require processing a large quantity of data. Current 4G networks standards suffer from some limitations such as those related to bandwidth, latency, and uniformity, especially when the data needs to be fed remotely. In this regard, 5G is likely to unlock the full potential of VR and AR technologies. 5G's significantly faster speeds and lower latency would help overcome these weaknesses.

参考译文

5G及应用

5G是最新一代蜂窝移动通信。5G性能目标在于高数据速率、低延时、低能耗、低成本、更大的系统容量和大规模设备连接。Release-15会在2019年4月完成5G规范的第一阶段,以适应早期的商业部署。Release-16的第二阶段将会在2020年4月完成,并提交ITU作为IMT-2020的候选技术。5G将会连接万物,为各行各业带来好处。5G会结合大数据、云计算、人工智能和许多其他创新技术,在下一个10年会加速信息黄金时代的到来。

随着物联网浪潮的到来,除了智能电话外越来越多的智能设备要接入移动互联网。下一代移动通信系统也称作第五代系统(5G),将会提供三种场景类型:第一种是增强型移动宽带eMBB,其目的是向以人为中心的用例提供宽带多媒体;第二种是超可靠低延时服务URLLC,其对延时(毫秒级)和可靠性(99.999%甚至更高)有严格的要求,超可靠低延时服务URLLC用于机器人远程控制或者触觉互联网应用;第三种是大规模机器类通信mMTC,主要用于连接大量的设备和发送低容量非延时敏感信息。相信从2010年到2020年总吞吐量将会增长1 000倍,连接设备数量增长到5 000亿。

为实现容量增长,5G的小区必须密集部署,大约是4G网络的40倍到50倍。由于新型空间接口技术,以及各种各样服务和终端的应用,一个典型的5G节点需要配置大约2 000个参数。5G规划还旨在降低延迟和降低能耗,以更好地实施物联网(IoT)。更具体地说,5G无线系统有8个高级功能:到现场端点的1~10 Gbit/s连接、1 ms的延迟、每单位面积1 000倍的带宽、10~100倍数量的连接设备、99.999%的可用性、100%的覆盖率、网络能耗降低90%和低功耗设备电池寿命长达10年。

5G网络通过使用更高的无线频率来实现更高的数据速率,使用毫米波频段在28 GHz至39 GHz之间,而之前的蜂窝网络使用的频率范围在700 MHz至3 GHz之间。一些供应商会使用低于6 GHz的第二低的微波频率范围,但这将不会有新频率的高速率。由于这些频率的带宽更丰富,5G网络将会使用更宽的频率通道与无线设备通信,最高可达400 MHz,而在4G LTE中则为20 MHz,每秒可传输更多的数据比特。5G采用正交频分复用OFDM调制,在这种调制方式中,多个载波在频道中传输,因此多个比特的信息可以同时并行传输。

毫米波被大气中的气体吸收,其传播范围比微波小,因此小区的覆盖范围受到限制;5G的小区范围有街区那么大,而以前蜂窝网络可能有好几英里宽。电波穿透建筑物的墙壁也有困难,需要许多天线来覆盖小区。毫米波天线比以前蜂窝网络中使用的大型天线要小,只有几英寸长,所以5G小区将被安装在电线杆和建筑物上的许多天线覆盖,而不是小区发射塔。

另一种提高数据速率的技术是大规模MIMO(多输入多输出)。每个小区将有多个天线与无线设备通信,每个天线通过一个单独的频率通道,由设备中的多个天线接收,因此多个比特流数据可以同时并行传输。在一种称为波束形成的技术中,基站计算机将不断计算无线电波到达每个无线设备的最佳路径，并将组织多个天线作为相控阵列一起工作，产生毫米波波束到达设备。相比于以前的蜂窝网络，5G网络小区规模更小、数量更多，使其网络基础设施每平方千米的覆盖建设成本更高。目前的部署限于城区，城区每个小区将有足够的用户提供较好的投资回报，而且人们怀疑这项技术是否会推广到农村地区。

5G与其他技术相结合，如物联网（IoT）、增强现实（AR）和虚拟现实（VR）、区块链、人工智能（AI），可能成为电子商务行业和市场的变革力量。

物联网

一项研究发现，全球70%的零售商已准备好采用物联网来改善消费者体验。电子商务活动可以通过多种方式从物联网中受益。例如，物联网可以轻松实时跟踪库存并更有效地管理库存。这样做可以减少人为错误。物联网还有助于减少浪费、控制成本并减少短缺。IoT设备产生了大量非结构化数据，并且产生的数据量增长速度是可用带宽的两倍。据估计到2020年，将需要一个至少是2016年水平的1 000倍的网络容量。需要处理的通信量也会以指数方式增加。

物联网是5G商用的基础，发展5G是为了给我们的生产和生活带来便利。而物联网就为5G提供了一个大展拳脚的舞台。在这个舞台上5G可以通过众多的物联网应用，如智慧农业、智慧物流、智能家居、车联网、智慧城市等，发挥出自己强大的作用。

虚拟现实和增强现实

VR和AR很可能成为电子商务行业和市场的推动力量。通过佩戴VR头盔，购物者可以立即发现自己在公司的虚拟商店中，在那里她可以四处“走动”寻找与在真实商店中完全一样的物品。例如，如果她想了解更多关于商店中珠宝新品的信息，她可以将注意力集中在该展品上，并查看做出购买决定所需的相关信息。如果她想购买，她可以付款或将其添加到购物车并查找其他商品。

AR和VR世界的复杂性和丰富性需要处理大量数据。当前的4G网络标准受到一些限制，例如与带宽、等待时间和一致性有关的限制，特别是当需要远程传送数据时。在这方面，5G可能会释放VR和AR技术的全部潜力。5G更快的速度和更低的延迟将有助于克服这些弱点。5G数据流的传输延迟大约是1 ms，这比人类可以注意到的时长要短得多。因此，5G几乎零传输延迟可能会丰富客户对AR和VR技术的体验。

区块链

用以促进电子商务活动的区块链复杂应用已经或正在开发。京东在其供应链管理系统和B2B电子商务中已经实施了区块链。2017年，随着牛肉制造商Kerchin成为其第一个供应链合作伙伴，该系统开始运行。该公司宣布，京东将于2017年10月底在其区块链上拥有超过10个品牌的酒精饮料、食品、茶叶和药品。

未来区块链最引人注目的用途之一可能是智能合约。在线供应商可以使用智能合约自动执行数字产品交付订单。在区块链上执行的智能合约中，5G可以在更有效地提供信息（例如，来自物联网设备）方面发挥关键作用。

AI

支持AI的设备已经在帮助消费者进行电子商务活动方面发挥着重要作用，如做出购买决策和跟踪产品。例如,虚拟助手正在改变消费者在线购买产品的方式。亚马逊的个人助理 Alexa 已经集成到亚马逊产品以及其他制造商的产品中。客户可以通过 eBay 的在线票务交换公司 StubHub 使用 Alexa 查找有关本地音乐会的信息。此外,他们还可以通过优步安排前往活动的交通,并订购 Domino's 的活动前晚餐。对订单状态可以实时跟踪。

5G 将极大地改善基于 AI 的设备的消费者体验。借助 5G,基于 AI 的设备可以更快地访问其他结构化和非结构化信息,并更好地了解环境。总体而言,人工智能支持的服务在其运营的各种环境和情况下将更加可靠。

作为 5G 技术的领导者,华为和主流芯片、终端和网络销售商一起完成了互操作开发测试。华为成为第一家公司发布了业界领先的 5G 商用芯片霸龙 5G01 和兼容 3GPP Release 15 的商用 CPE。华为也是能够提供端到端商用解决方案的唯一的销售商,正积极地促进 5G 产业链的成熟和商业应用。

通过部署 5G 网络,例如使用物联网设备,可以克服当前电子商务发展中的一些挑战。处理物联网设备的指数增长将是一个不小的壮举。然而,当前的 4G 网络无法处理来自 IoT 设备的所有数据。5G 更高的数据传输和处理速度将解决这一问题。具体而言,5G(与 AI、VR、AR 和其他技术相结合)将在改变电子商务行业和市场方面发挥强大的作用。这种组合可以带来丰富的电子商务生态系统和更好的客户体验。

UNIT 6

PASSAGE

The Vision of 6G[1]

6G, also an extension of 5G, is the standard of sixth generation mobile communication. 5G is being driven by consumers' growing demand for traffic and the demand for productivity in vertical industries. In other words, it is essentially commercial demand that drives the development of 5G.[2] 6G is driven by the commercial and social needs. A 6G base station can be connected to hundreds or even thousands of wireless connections at the same time, and its capacity can be up to 1 000 times that of a 5G base station.

On March 19,2019, the Federal Communications Commission(FCC) decided to open up the "terahertz" spectrum for future 6G network services or innovators to test the 6G technology. On November 3, China announced the establishment of a national working group and a group of experts on the promotion of 6G technology research and development, marking the official launch of China's 6G technology research and development.[3]

There are five application scenarios supported by 6G communication technologies.

1. eMBB-Plus (Enhanced Mobile Broadband Plus). It is the successor of eMBB of 5G and it should enhance the performance of mobile communications in terms of interference, handover, data transmission, data processing and interoperability but also guaranteeing a high level of security and privacy.

2. BigCom (Big Communications). It aims at extending communication services to remote areas by guaranteeing network coverage and adequate quality of service everywhere.

3. SURLLC (Secure Ultra-Reliable Low-Latency Communications). It should provide higher reliability than 5G URLLC and lower latency than 5G mMTC, including security aspects.

4. 3D-InteCom (Three-Dimensional Integrated Communications). It addresses 3D network analysis, planning and optimization, that takes into account the heights of network nodes (e.g. satellite, unmanned aerial vehicle, underwater communications, full-dimensional MIMO architectures).

5. UCDC (Unconventional Data Communications). It includes new communication prototypes and paradigms that may range (but are not limited to) from holographic and tactile Internet to human-bond communications.

In summary, the main requirements for 6G are the following: high-speed and high-capacity communications, extreme coverage extension, low power consumption, low latency, high reliable communications, massive connectivity and sensing and high security and privacy.

The extension of the cyberspace and the adoption of big data and AI technologies ask for enhancements of security and privacy. On the other hand, although the improvement of the spectral efficiency in 6G would not be on a large scale due to the achievements already achieved by 5G technology, high-speed and high-capacity communications will be the basis for the provision of advanced virtual/augmented user experience. Moreover, novel energy efficiency solutions, including low power consumption devices, are expected to be defined in compliance with the green communication paradigm.[4] Finally, new network architectures are required to cope with the dynamicity of network topology and to ensure high affordability and full customization based on advanced AI implementations.

A wide range of technologies can be considered for 6G to meet the challenges to come and many of these are suitable to be jointly evaluated in research collaborations. Here, we highlight a selection of components that we see as potential technological enablers for the new capabilities we foresee 6G to have.[5]

6G will use the THz frequency band (300 GHz to 3 000 GHz) and the 6G network will be "densified" to an unprecedented level, at which we will be surrounded fully by small base stations.[6] The frequency of the 6G signal is already at the level of terahertz, which is close to the spectrum of the molecular rotational energy level.[7] It is easily absorbed by the water molecules in the air, so it cannot travel as far in space as the 5G signal. Therefore, the 6G signal needs the "relay" of more base stations.

6G will use "spatial muliplexing technology". The 6G base station will be able to connect hundreds or even thousands of wireless connections at the same time, and its capacity will be 1 000 times that of a 5G base station. The higher frequency means the greater propagation loss. The closer the coverage distance is, the weaker the diffraction ability is. 6G and 5G face the same problem. However, 5G solves this kind of problems using the key technology of massive MIMO and beamforming.[8] The traffic of transmitted data can be increased under the configuration of multiple MIMO antennas. The advantage of this technology is that it can increase channel capacity and improve spectrum utilization without consuming extra bandwidth and transmitting power.[9] MIMO technology is used in 5G to improve spectrum utilization. Since 6G is located in a higher frequency band, further development of MIMO is likely to provide key technical support for 6G in the future.

Future networks are expected to utilize AI agents for multiple functions, including optimization of radio interface, network management automation (such as optimization of parameters, handling of alarms, self healing) and orchestration.[10] To enable this, precise and timely data needs to be readily available at the right places in the network, so that the intelligent agents can access the data. At the same time, the networks should stream the needed data from different domains and sources in an efficient manner to avoid the transmission and storage of massive amounts of data that may never be utilized over network management interfaces.[11] AI algorithms should be deployed and trained at different levels of the network: management layer, core, radio base stations, and as well as in the UE, possibly with the assistance of the network.

5G introduced a service-based architecture in the core network (CN) level, and in 6G service-based network design can be taken one step further to encompass also the radio access network (RAN) and possible elements of UE. It should open up for efficient cloud implementation of all network functions, reusing a common framework of service discovery, data storage, etc., and avoiding duplicated functionality in RAN and CN, as well as unnecessary proxy functions. This ensures an efficient setup where features can be flexibly added by combining network functions in new ways or introducing additional network functions with minimal impediment.[12]

With the rapid development of network technology, display technology, sensing and imaging equipment, and low-power processor, in the 6G era, VR, AR, and MR techniques should be integrated seamlessly with human sense through wearable displays, sensors, network, replacing today's smart phones and becoming the main tools of human entertainment, life and work.[13]

In the age of 6G, with the continuous development of high-resolution imaging, sensing, wearable display, mobile robot, processor and wireless network technology, remote holography will become a reality. Remote holography transmits 3D holograms of people in different places to the same location through real-time capture, transmission and rendering technology, enabling people to communicate as if they were sitting face to face.[14] Remote holography is not limited to the emotional communication between people, it will be widely used in distance education, collaborative design, telemedicine, telecommuting, remote training and other fields.

By 2030 and beyond, millions of self-driving cars connected around the world will drive collaboratively on a 6G network to make transportation and logistics as efficient as possible. The benefits of networked self-driving cars include reducing traffic congestion, reducing tailpipe emissions, improving transportation efficiency, and improving driving safety.

Although edge clouds are already used widely in the 5G era, it is believed that they will be unprecedentedly prosperous in the 6G era, further spawning a flood of new services.[15] The cloud capacity of edge cloud sinks locally, bringing cloud computing power and applications closer to the user side, thereby reducing latency and network load.[16] Just as today's cloud applications are booming, so the edge cloud of the future is bound to spawn a flood of localized real-time applications and services.

With the development of 5G, the future mobile communication network will enter all walks of life. Most vertical industries need location services, such as asset tracking, precision marketing, transportation and logistics, AR, healthcare, etc., but traditional satellite positioning is not accurate in urban and indoor scenes.[17] In the 6G era, the 3D beamforming technology can realize accurate positioning at the centimeter level, and its integration with the constantly developing induction, imaging and other technologies will promote a large number of new applications.

NEW WORDS AND PHRASES

vision n. 视力,美景,想象力
traffic n. 交通,运输,通信量
productivity n. 生产力,生产能力
vertical a. 垂直的,直立的
commercial a. 商业的,商务的
terahertz n. 太赫兹
announce v. 宣布,宣告
promotion n. 提升,促进
successor n. 继承者,后续的事情
interference n. 干扰,干涉
handover n. 切换,换手
interoperability n. 互操作性,互通
guarantee vt. 保证,担保
reliability n. 可靠性,可靠度
address vt. 设法解决,提出
optimization n. 最佳化,优化
aerial a. 空中的,航空的
architecture n. 建筑风格,建筑样式,架构
prototype n. 原型,标准,模范
paradigm n. 范式,范例
holographic a. 全息的
tactile a. 触觉的,有触觉的
consumption n. 消费,消耗
massive a. 大量的,巨大的
augment v. 增加,增大
dynamicity n. 动态,动态性
topology n. 拓扑,拓扑学

customization	*n*. 定制，客户化
collaboration	*n*. 合作，协作
densify	*vt*. 致密，增密
unprecedented	*a*. 空前的，无前例的
molecular	*a*. 分子的，由分子组成的
propagation	*n*. 传播，繁殖
diffraction	*n*. 衍射，绕射
beam	*n*. 横梁，光纤，电波
utilization	*n*. 利用，使用
parameter	*n*. 参数，参量
orchestration	*n*. 管弦乐编曲，和谐的结合，编排，配器
agent	*n*. 代理人，代理商
algorithm	*n*. 算法，运算法则
encompass	*vt*. 包含，包围
framework	*nt*. 框架，结构
duplicate	*vt*. 复制，重复
impediment	*n*. 口吃，妨碍，阻碍
seamless	*a*. 无缝的，如缝合线的
wearable	*a*. 可穿戴的，穿戴舒适的
resolution	*n*. 分辨率，决议
robot	*n*. 机器人，机械人
capture	*vt*. 俘获，夺得
render	*v*. 致使，提供
emotional	*a*. 情绪的，易激动的
telemedicine	*n*. 远程医疗，远程医学
logistics	*n*. 后勤，物流
congestion	*n*. 拥挤，拥塞
tailpipe	*n*. 排气管
emission	*n*. 发射，散发
prosperous	*a*. 繁荣的，兴旺的
edge	*n*. 边缘，缘，刀刃

NOTES

1. 6G 远景。本文对 6G 技术及应用远景进行了说明。

2. It is … that 为强调句式。本句译文：换句话说，本质上正是商业需求驱动了 5G 的发展。

3. marking the official launch of China's 6G technology research and development 为现在分词短语,表示伴随情况;promotion 本身是名词,翻译时按动词来翻译。本句译文:11 月 3 日,为促进 6G 技术的研究和开发,中国宣布成立国家 6G 技术研发推进工作组和总体专家组,这标志着我国 6G 技术研发工作正式启动。

4. 本句是被动句; including low power consumption devices 为分词短语,表示伴随情况;in compliance with,按照。本句译文:此外,新型的能源效率解决方案,包括低功耗设备,预计将确定符合绿色通信范例。

5. that 引导定语从句,修饰 components;we foresee 6G to have 是省略 that 的定语从句。本句译文:在这里,我们强调了一些技术组成,我们认为这些技术是我们预计 6G 将具有的新功能的潜在技术成因。

6. 由 and 连接两个并列句;at which we will be surrounded fully by small base stations 为定语从句,介词前置。本句译文:6G 将使用太赫兹(THz)频段(300～3 000 GHz),且 6G 网络的"致密化"程度也将达到前所未有的水平,届时,我们的周围将充满小基站。

7. which is close to the spectrum of the molecular rotational energy level 为非限定性定语从句;close to,接近于。本句译文:6G 信号的频率已经在太赫兹级别,而这个频率已经接近分子转动能级的光谱了。

8. using the key technology of massive MIMO and beamforming 为现在分词短语,表示伴随情况;beamforming,波束形成。本句译文:5G 则是通过大规模 MIMO 和波束形成关键技术来解决此类问题的。

9. that 引导表语从句,做系动词 is 的表语; consuming extra bandwidth and transmitting power 动名词短语做介词 without 的宾语。本句译文:这种技术的好处就是,它能够在不占用额外带宽、消耗额外发射功率的情况下增加信道容量,提高频谱利用率。

10. 本句是被动句;to utilize AI agents for multiple functions 不定式短语表示目的;including 引导现在分词短语,表示伴随情况。本句译文:未来的网络有望利用 AI 代理实现多种功能,包括无线接口优化、网络管理自动化(如参数优化、报警处理、自愈)和业务编排。

11. to avoid the transmission and storage of massive amounts of data 为不定式短语,表示目的;that may never be utilized over network management interfaces 为定语从句,修饰 data。本句译文:与此同时,网络应该以一种有效的方式从不同的域和源传输所需的数据,以避免传输和存储可能永远不会在网络管理接口上使用的大量数据。

12. where 引导定语从句,修饰 setup; combining network functions in new ways or introducing additional network functions with minimal impediment 动名词短语做介词 by 的宾语。本句译文:这确保了一种高效的设置,以新的方式组合网络功能或以最小的障碍引入额外的网络功能,可以灵活地添加特性。

13. replacing today's smart phones and becoming the main tools of human entertainment, life and work 两个并列的现在分词短语,表示伴随情况。本句译文:随着网络技术、显示技术、传感和成像设备以及低功耗处理器的飞速发展,到了 6G 时代,VR、

AR 和 MR 技术将通过可穿戴显示器、传感设备、网络与人类感官无缝集成，替代今天的智能手机，成为人类娱乐、生活和工作的主要工具。

14. enabling people to communicate 为现在分词短语，表示伴随情况；as if，好像，似乎；As if they were sitting face to face 为虚拟语句。本句译文：远程全息通过实时捕获、传输和渲染技术，将身处不同地方的人的 3D 全息影像传送到同一位置，使大家如面对面坐在一起一样交流沟通。

15. it is … that，it 是形式主语；spawning a flood of new services 为现在分词短语，表示伴随情况；a flood of，大量的，一大批。本句译文：尽管在 5G 时代边缘云已经得到普遍应用，但我们相信其在 6G 时代将空前繁荣，并进一步催生海量新服务。

16. bringing cloud computing power and applications closer to the user side 为现在分词短语，表示伴随情况；reducing latency and network load 为现在分词短语，表示伴随情况。本句译文：边缘云将云端能力下沉到本地，让云计算能力和应用更接近用户侧，从而减少时延和网络负荷。

17. location services，定位服务；such as，例如；satellite positioning，卫星定位。本句译文：大部分的垂直行业都需要定位服务，比如资产跟踪、精准营销、运输和物流、AR、医疗保健等应用，但传统卫星定位方法在城市和室内场景中并不精准。

EXERCISES

一、请将下述词组译成英文

1. 5G 基站　2. 未来的 6G 网络服务　3. 增强型移动宽带
4. 保证网络覆盖　5. 无人航空器　6. 低功率消耗
7. 安全和隐私　8. 太赫兹水平　9. 波束形成
10. 增加信道容量　11. 自动驾驶汽车　12. 改进的驾驶安全
13. 定位服务　14. 无线接入网　15. 高分辨率成像
16. 改善频谱使用　17. 无线接口优化　18. 更大的传输损耗
19. 基于服务的架构　20. 传感和成像设备　21. 未来的边缘云
22. 实现精确定位

二、请将下述词组译成中文

1. the growing demand of consumers for services
2. the demand for productivity in vertical industries
3. the commercial and social needs
4. thousands of wireless connections
5. to open up the “terahertz” spectrum
6. the promotion of 6G technology research and development
7. the application scenarios supported by 6G
8. the performance of mobile communications
9. a high level of security and privacy
10. the holographic and tactile Internet

11. the improvement of the spectral efficiency
12. high-speed and high-capacity communications
13. the advanced virtual/augmented user experience
14. the new network architectures
15. the dynamicity of network topology
16. the full customization based on advanced AI
17. to meet the challenges to come
18. the spatial muliplexing technology
19. the configuration of multiple MIMO antennas
20. at different levels of the network
21. the efficient cloud implementation
22. to introduce additional network functions
23. the rapid development of network technology
24. the main tools of human entertainment
25. to transmit 3D holograms of people in different places

三、选择合适的答案填空

1. 5G is being driven by consumers' ________ demand for traffic and the demand for productivity in vertical industries.

A. grow B. to grow
C. growing D. growed

2. On March 19, 2019, the Federal Communications Commission (FCC) decided ________ up the "terahertz" spectrum for future 6G network services or innovators to test the 6G technology.

A. to open B. open
C. opening D. opened

3. Moreover, novel energy efficiency solutions, ________ low power consumption devices, are expected to be defined in compliance with the green communication paradigm.

A. including B. included
C. to include D. include

4. A wide range of technologies can be considered for 6G to meet the challenges to come and many of these are suitable to be jointly ________ in research collaborations.

A. evaluate B. to evaluate
C. evaluating D. evaluated

5. The traffic of ________ data can be increased under the configuration of multiple MIMO antennas.

A. transmitting B. transmitted
C. to transmit D. transmit

6. Just as today's cloud applications are booming, so the edge cloud of the future is bound ________ a flood of localized real-time applications and services.

A. spawn　B. spawning

C. to spawn　D. spawned

7. The benefits of networked self-driving cars include reduced traffic congestion, ________ tailpipe emissions, improved transportation efficiency, and improved driving safety.

A. reduce　B. reduced

C. to reduce　D. reducing

四、根据课文内容选择答案

1. A 6G base station can be connected to hundreds or even ________ at the same time, and its capacity can be up to 1 000 times that of a 5G base station.

A. thousands of wireless connections

B. the demand for productivity

C. a group of experts on the promotion

D. the performance of mobile communications

2. The extension of the cyberspace and the adoption of Big Data and AI technologies ask for ________.

A. high-speed and high-capacity communications

B. low power consumption devices

C. communication prototypes and paradigms

D. enhancements of security and privacy

3. Moreover, novel energy efficiency solutions, including ________, are expected to be defined in compliance with the green communication paradigm.

A. the adoption of Big Data

B. low power consumption devices

C. high reliable communications

D. 3D network analysis

4. ________ can be considered for 6G to meet the challenges to come and many of these are suitable to be jointly evaluated in research collaborations.

A. A wide range of technologies

B. The dynamicity of network topology

C. The heights of network nodes

D. A high level of security

5. It is easily absorbed by the water molecules in the air, so it cannot travel as far in space as the 5G signal. Therefore, the 6G signal needs the "relay" ________.

A. of the molecular rotational energy

B. of network topology

C. of more base stations

D. of advanced virtual/augmented user experience

6. MIMO technology is used in 5G to ________.

A. solve this kind of problems

B. improve spectrum utilization

C. ensure high affordability

D. mean the greater propagation loss

7. Future networks are expected to utilize AI agents for multiple functions, including ________, network management automation (such as optimization of parameters, handling of alarms, self healing) and orchestration.

A. different levels of the network

B. massive amounts of data

C. optimization of radio interface

D. possible elements of UE

五、请将下列短文译成中文

1. 5G is being driven by consumers' growing demand for traffic and the demand for productivity in vertical industries. In other words, it is essentially commercial demand that drives the development of 5G.

2. On November 3, China announced the establishment of a national working group and a group of experts on the promotion of 6G technology research and development, marking the official launch of China's 6G technology research and development. There are five application scenarios supported by 6G communication technologies.

3. In summary, the main requirements for 6G are the following: high-speed and high-capacity communications, extreme coverage extension, low power consumption, low latency, high reliable communications, massive connectivity and sensing and high security and privacy.

4. On the other hand, although the improvement of the spectral efficiency in 6G would not be on a large scale due to the achievements already achieved by 5G technology, high-speed and high-capacity communications will be the basis for the provision of advanced virtual/augmented user experience.

5. A wide range of technologies can be considered for 6G to meet the challenges to come and many of these are suitable to be jointly evaluated in research collaborations. Here, we highlight a selection of components that we see as potential technological enablers for the new capabilities we foresee 6G to have.

6. The frequency of the 6G signal is already at the level of terahertz, which is close to the spectrum of the molecular rotational energy level. It is easily absorbed by the water molecules in the air, so it cannot travel as far in space as the 5G signal. Therefore, the 6G signal needs the "relay" of more base stations.

7. 6G will use "spatial muliplexing technology". The 6G base station will be able to connect hundreds or even thousands of wireless connections at the same time, and its capacity will be 1 000 times that of a 5G base station. The higher frequency means the greater propagation loss.

8. With the development of 5G, the future mobile communication network will enter all walks of life. Most vertical industries need location services, such as asset tracking, precision marketing, transportation and logistics, AR, healthcare, etc., but traditional satellite positioning is not accurate in urban and indoor scenes.

参考译文

6G愿景

6G,也是5G之后的延伸,即第六代移动通信标准。5G的驱动力来源于消费者不断增长的流量需求,以及垂直行业的生产力需求。换句话说,本质上是商业需求驱动了5G的发展。6G的驱动力来源于商业需求和社会需求。6G基站可同时接入数百个甚至数千个无线连接,其容量可达5G基站的1 000倍。

2019年3月19日,美国联邦通信委员会(FCC)决定开放面向未来6G网络服务的"太赫兹"频谱,用于创新者开展6G技术试验。11月3日,中国宣布成立国家6G技术研发推进工作组和总体专家组,这标志着我国6G技术研发工作正式启动。

6G通信技术支持5种应用场景。

1. eMBB-Plus(增强的移动宽带+)。它是5G的eMBB的继承者,将会提高移动通信在干扰、切换、数据传输、数据处理和互操作性方面的性能,同时具有高度安全和隐私保护。

2. BigCom(大通信)。它的目标是在确保网络覆盖和各地的服务质量的同时将通信服务扩展到偏远地区。

3. SURLLC(安全的超可靠低时延通信)。它将提供比5G URLLC更高的可靠性和比5G mMTC更低的延迟,包括安全方面。

4. 3D-InteCom(三维综合通信)。它处理三维网络分析、规划和优化,考虑网络节点的高度(如卫星、无人飞行器、水下通信、全维MIMO架构)。

5. UCDC(非传统数据通信)。它包括了新的通信原型和范例,可能包括(但不限于)从全息和触觉互联网到人际联系通信。

综上所述,6G的主要需求是:高速大容量通信、最大程度扩展覆盖、低功耗、低延迟、高可靠通信、海量连通性和传感、高安全性和私密性。

网络空间的扩展、大数据和人工智能技术的应用,都要求增强安全性和隐私保护。另一方面,尽管鉴于5G技术已经取得的成就,6G的频谱效率不会有大的提升,但高速、大容量的通信将是提供先进的虚拟/增强用户体验的基础。此外,新型的能源效率解决方案,包括低功耗设备,预计将定义符合绿色通信范例。最后,需要新的网络架构来应对网络拓扑的动态性,并确保高可购性和基于高级AI实现的完全定制。

6G可以考虑使用多种技术来应对未来的挑战,其中许多技术适合在研究协作中进行联合评估。在这里,我们强调了一些技术组成,我们认为这些技术是我们预计6G将具有的新功能的潜在技术成因。

6G将使用太赫兹(THz)频段(300～3 000 GHz),且6G网络的"致密化"程度也将达到前所未有的水平,届时,我们的周围将充满小基站。6G信号的频率已经在太赫兹级别,而这个频率已经接近分子转动能级的光谱了。6G信号很容易被空气中的水分子吸收,所以在空间中传播的距离不像5G信号那么远,因此6G需要更多的基站"接力"。

6G将使用"空间复用技术"。6G基站将可同时接入数百个甚至数千个无线连接,其容量将可达到5G基站的1 000倍。更高的频率意味着更大的传输损耗。频率越高,传播损耗越大。覆盖距离越近,绕射能力越弱。6G和5G面临相同的问题。5G则是通过大规模MIMO和波束形成关键技术来解决此类问题的。在MIMO多副天线的配置下可以提高传输数据量。这种技术的好处是,它能够在不占用额外带宽、消耗额外发射功率的情况下增加信道容量,提高频谱利用率。5G采用MIMO技术提高频谱利用率。而6G所处的频段更高,MIMO未来的进一步发展很有可能为6G提供关键的技术支持。

未来的网络有望利用AI代理实现多种功能,包括无线接口优化、网络管理自动化(如参数优化、报警处理、自愈)和业务编排。为了实现这一点,需要在网络中的正确位置随时提供精确和及时的数据,以便智能代理能够访问数据。与此同时,网络应该以一种有效的方式从不同的域和源传输所需的数据,以避免传输和存储可能永远不会在网络管理接口上使用的大量数据。人工智能算法应该在网络的不同层次上进行部署和培训:管理层、核心、无线电基站,以及UE,可能的话,在网络的帮助下。

5G在核心网络(CN)层引入了一个基于服务的架构,而在6G中,基于服务的网络设计可以进一步包含无线接入网(RAN)和可能的UE元素。它应该开放所有网络功能的有效云实现,重用服务发现的通用架构、数据存储等,避免RAN和CN中的重复功能,以及不必要的代理功能。这确保了一种高效的设置,以新的方式组合网络功能或以最小的障碍引入额外的网络功能,可以灵活地添加特性。

随着网络技术、显示技术、传感和成像设备以及低功耗处理器的飞速发展,到了6G时代,VR、AR和MR技术将通过可穿戴显示器、传感设备、网络与人类感官无缝集成,替代今天的智能手机,成为人类娱乐、生活和工作的主要工具。

到了6G时代,随着高分辨率成像、传感、可穿戴显示器、移动机器人、处理器和无线网络技术的不断发展,远程全息将成为现实。远程全息通过实时捕获、传输和渲染技术,将身处不同地方的人的3D全息影像传送到同一位置,使大家如面对面坐在一起一样交流沟通。远程全息不仅限于人与人之间的情感沟通,它将广泛应用于远程教育、协作设计、远程医疗、远程办公、远程培训等领域。

2030年及以后,全球数百万联网的自动驾驶汽车将在6G网络下协同行驶,以使运输和物流尽可能更高效。联网自动驾驶汽车的好处在于可减少交通拥堵、减少尾气排放、提升运输效率,以及提高驾驶安全性。

尽管在5G时代边缘云已经得到普遍应用,但我们相信其在6G时代将空前繁荣,并进一步催生海量新服务。边缘云将云端能力下沉到本地,让云计算能力和应用更接近用户侧,从而减少时延和网络负荷。就像今天的云应用繁荣生长一样,相信未来边缘云也必将催生海量的本地化即时应用和服务。

随着5G发展,未来的移动通信网络将进入各行各业。大部分的垂直行业都需要定位服务,比如资产跟踪、精准营销、运输和物流、AR、医疗保健等应用,但传统卫星定位方法在城市和室内场景中并不精准。6G时代的3D波束形成技术可实现厘米级的精准定位,其与不断发展的感应、成像等技术集成,将催生海量新应用。

UNIT 7

PASSAGE

Cloud Computing[1]

The cloud computing is one kind of emerging business computing model. It distributes the duty of calculation to the resource pool which the massive computers constitute, enables each kind of application system according to need to gain the computation strength, the storage space and all kinds of software service.[2] This kind of resource pool is called "the cloud". "The cloud" is virtual computation resources that can maintain and manage itself, usually for some large-scale server cluster, including calculating server, storage server, the broad band resources and so on.[3] The cloud computing will concentrate all computation resources, and can be managed automatically through the software without intervention. The users who use cloud computing will not be worried by the tedious details and concentrate his own business. Cloud computing is a recent trend in IT that moves computing and data away from desktop and portable PCs into large data centers.[4]

The major cloud providers such as Google, Microsoft and Amazon have built and are working on building the world's largest data centers across the United States and elsewhere. Each data center includes hundreds of thousands of computer servers, cooling equipment and substation power transformers.

With the large scale proliferation of the Internet around the world, applications can now be delivered as services over the Internet.[5] The main goal of cloud computing is to make a better use of distributed resources, combine them to achieve higher throughput and be able to solve large scale computation problems. Cloud computing has super computing power. Thousands of computers form a super server in cloud services which provide users with powerful computing and data processing capacity that is hard to realize for a personal computer.[6] As a result this reduces the overall cost. In cloud computing, users access the data, applications or any other services with the help of a browser regardless of the device used and the user's location.

Cloud computing has been envisioned as the next-generation computing model for its major advantages in on-demand self-service, ubiquitous network access, location independent resource pooling and transference of risk.[7] On-demand self-service allows users to obtain, configure and deploy cloud services themselves using cloud service catalogues, without requiring the assistance of IT. Cloud computing is the latest developments of computing models after distributed computing, parallel processing and grid computing. The definition of cloud computing provided by National Institute of Standards and Technology (NIST) says that: "Cloud computing is a model for enabling convenient, on-demand network access to a shared pool of configurable computing resources (e.g., networks, servers, storage applications and services) that can be rapidly provisioned and released with minimal management effort or service provider interaction".[8]

"Cloud computing" simply means "Internet computing", generally the Internet is seen as collection of clouds; thus the word cloud computing can be defined as utilizing the Internet to provide technology enabled services to the people and organizations.[9] Cloud computing enables consumers to access resources online through the Internet, from anywhere at any time without worrying about technical/physical management and maintenance issues of the original resources.[10] Google Apps is the paramount example of cloud computing, it enables users to access services via the browser and deployed on millions of machines over the Internet. Cloud computing is cheaper than other computing models; zero maintenance cost is involved since the service provider is responsible for the availability of services and clients are free from maintenance and management problems of the resource machines.[11] The principal service models being deployed are: Software as a Service (SaaS), Platform as a Service (PaaS), Infrastructure as a Service (IaaS).

Fig. 7-1 shows the different layers of cloud computing architecture.

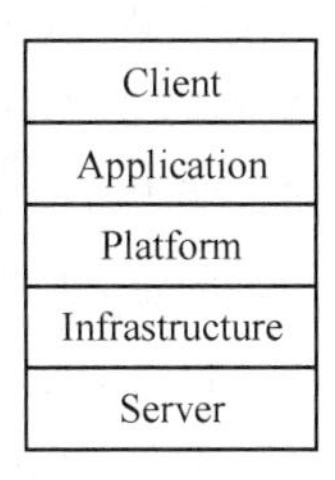

Fig. 7-1 Layers of Cloud Computing Architecture

A cloud client consists of computer hardware and/or computer software which relies on cloud computing for application delivery.[12]

A cloud application delivers "Software as a Service (SaaS)" over the Internet, thus eliminating the need to install and run the application on the users system. Softwares are provided as a service to the consumers according to their requirement, which enable consumers to use the services that are hosted on the cloud server.[13] Google Apps is the most widely used SaaS.

Platform services "Platform as a Service (PaaS)" provide a computing platform using the cloud infrastructure.[14] Clients are provided platforms access, which enables them to put their own customized software's and other applications on the clouds. Through this service developers can get a hold of all the systems and environments required for the life cycle of software, such as its developing, testing, deploying and hosting of web applications.

Infrastructure services "Infrastructure as a Service (IaaS)" provide the required infrastructure as a service. Rent processing, storage, network capacity, and other basic computing resources are granted, enabling consumers to manage the operating systems, applications, storage, and network connectivity. The client need not purchase the required servers, data center or the network resources. Also the key advantage here is that customers need to pay only for the time duration they use the service.[15] As a result customers can achieve a much faster service delivery with less cost.

NEW WORDS AND PHRASES

emerging	a. 新兴的,出现的
distribute	vt. 分配,散布
massive	a. 大量的,巨大的
pool	n. 联营,水池
virtual	a. 虚拟的
computation	n. 计算,估计
concentrate	v. 集中,浓缩
automatically	ad. 自动地,不经思索地
intervene	vi. 干涉,干预
tedious	a. 沉闷的,单调乏味的
substation	n. 变电站,变电所
proliferation	n. 增殖,扩散
capacity	n. 容量,能力
regardless	a. 无论,不管
envision	vt. 想象,预想
ubiquitous	a. 普遍存在的,无所不在的
transference	n. 转移,转让
parallel	a. 平行的,类似的
grid	n. 格子,网格
definition	n. 定义,清晰度
configurable	a. 可配置的,结构的
interaction	n. 互动,交互作用
online	a. 联机的,在线的
maintenance	n. 维护,维修
original	a. 原始的,最初的

paramount	*a*. 最重要的,主要的
browser	*n*. 浏览器,阅读器
responsible	*a*. 负责的,有责任的
deploy	*v*. 部署,展开
diagram	*n*. 图表,图解
architecture	*n*. 建筑学,建筑风格
eliminate	*vt*. 消除,排除
host	*v*. 主持,托管
platform	*n*. 平台,月台
customize	*vt*. 定制,自定义
environment	*n*. 环境,外界
grant	*v*. 同意,允许
connectivity	*n*. 连接,联通,连接性

NOTES

1. 本篇课文涉及计算机和互联网技术,题目为:云计算。

2. 本句主句有两个谓语 distributes 和 enables; which the massive computers constitute 限定性定语从句,修饰 pool; according to,根据; to gain the computation strength 不定式短语表示目的。本句可译成:它将计算任务分配给由大量计算机组成的资源池,并根据需要使每种应用系统获得计算能力、存储空间和各种软件服务。

3. that can maintain and manage itself 定语从句,修饰 resources; including calculating server, storage server, the broad band resources 为现在分词短语,说明 server cluster;and so on,等。本句可译成:"云"是能够自我维护和自我管理的虚拟计算资源,通常是针对大规模的服务器群,包括计算服务器、存储服务器、宽带资源等。

4. that moves computing and data away from desktop and portable PCs into large data centers 定语从句,修饰前面句子;from ... into,从……到。本句译成:云计算是信息技术的新趋势,它将计算及数据从桌面和手提电脑上转到大型数据中心。

5. 介词短语 with 结构表示伴随情况;本句是被动句,可以按主动句翻译。本句译成:随着互联网在全世界的大规模普及,应用业务可以通过互联网来提供。

6. which provided users with powerful computing and data processing capacity 引导限定性定语从句,修饰 services;that is hard to realize 定语从句修饰 capacity。本句译成:在云计算服务中,数千的计算机构成超级服务器,由这些服务器向用户提供强有力的数据计算和处理能力,这些能力是个人计算机难以实现的。

7. 本句是被动句的完成时态;for its major advantages 介词短语表示原因。本句译成:云计算因其按需自我服务、无所不在的网络接入、与位置无关的资源池以及风险转移的主要优势而被看成是下一代计算模式。

8. provided by National Institute of Standard and Technology 过去分词短语做后置

定语,修饰 computing;that can be rapidly provisioned and released 定语从句,修饰 resources。本句译成:美国国家标准与技术研究院提供的云计算的定义,云计算是一种模型,它可以实现随时随地、便捷地、随需应变地从可配置计算资源共享池中获取所需的资源(例如,网络、服务器、存储应用及服务),资源能够快速供应并释放,使管理资源的工作量和与服务提供商的交互减小到最低限度。

9. 本句有两个被动句;utilizing 做介词 as 的宾语;to provide 动词不定式表示目的。本句译成:云计算就意味着互联网计算,通常互联网被看成是云的集合,这样云计算可以定义为使用互联网向个人和组织机构提供服务的技术。

10. without worrying about 介词短语表示伴随情况。本句译成:云计算使得顾客能够随时随地通过互联网接入在线资源,而不必担心最初资源的管理和维护。

11. cheaper than 为比较级;is involved,被动结构;since 引导原因状语从句。本句译成:云计算比其他计算模型更便宜,维护成本为零,因为服务提供商负责服务的有效使用而顾客不受资源设备管理和维护的影响。

12. consist of,由……组成;which relies on cloud computing 定语从句,修饰 software;for application delivery 介词短语表示目的。本句译成:云客户由依赖提供云计算服务的计算机硬件和软件组成。

13. according to,根据;which enable consumers to use the services 非限定性定语从句,修饰前面句子;that are hosted on the cloud server 定语从句,修饰 services。本句译成:根据客户的需要,软件作为一种服务提供给客户,使用户能够使用托管在云服务器的服务。

14. using the cloud infrastructure 现在分词做后置定语。本句译成:平台服务"平台即服务(PaaS)"提供使用云结构的计算平台。

15. that customers need to pay 为表语从句;they use the service 为省略 that 的定语从句,修饰 duration。本句译成:主要的优势在于客户只需对使用服务的这段时间付费。

EXERCISES

一、请将下述词组译成英文

1. 云计算
2. 商业计算模型
3. 各种软件服务
4. 虚拟计算资源
5. 大型数据中心
6. 更好地使用分布资源
7. 超级计算能力
8. 降低整个费用
9. 借助浏览器的帮助
10. 无所不在的网络接入
11. 云计算的定义
12. 主要的服务模型
13. 云计算最突出的例子
14. 云计算结构分层
15. 提供计算平台
16. 计算平台
17. 云结构
18. 基本的计算资源

二、请将下述词组译成中文

1. to gain the computation strength
2. this kind of resource pool
3. some large-scale server cluster
4. to manage automatically through the software

5. the users who use cloud computing
6. the world's largest data centers
7. the major cloud providers
8. with the large scale proliferation of the Internet
9. to solve large scale computation problems
10. with the help of a browser
11. the next-generation computing model
12. a shared pool of configurable computing resources
13. to access resources online through the Internet
14. to install and run the application on the users system
15. to use the services that are hosted on the cloud server
16. the life cycle of software
17. to achieve a much faster service

三、选择合适的答案填空

1. This kind of resource pool is ________ "the cloud".

A. called　　B. call
C. calling　　D. to call

2. The cloud computing will concentrate all computation resources, and can be ________ automatically through the software without intervene.

A. manage　　B. to manage
C. managing　　D. managed

3. The main goal of cloud computing is ________ a better use of distributed resources, combine them to achieve higher throughput and be able to solve large scale computation problems.

A. making　　B. to make
C. made　　D. make

4. A cloud application delivers "Software as a Service (SaaS)" over the Internet, thus ________ the need to install and run the application on the users system.

A. to eliminate　　B. eliminating
C. eliminated　　D. eliminated

5. Platform services "Platform as a Service (PaaS)" provide a computing platform ________ the cloud infrastructure.

A. using　　B. to use
C. used　　D. use

6. As a result this reduces the overall cost. In cloud computing, users access the data, applications or any other services with the help of a browser regardless of the device ________ and the user's location.

A. using　　B. to use

C. used　　　　D. use

四、根据课文内容选择正确答案

1. This kind of ________ is called "the cloud".

A. resource pool　　　　B. storage server

C. tedious detail　　　　D. cooling equipment

2. The users who use cloud computing will not be worried by the tedious details and ________.

A. concentrate all computation resources

B. solve large scale computation

C. concentrate his own business

D. achieve higher throughput

3. The major cloud providers such as Google, Microsoft and Amazon have built and are working on ________ across the United States and elsewhere.

A. emerging business computing model

B. utilizing the Internet to provide technology

C. worrying about technical/physical management

D. building the world's largest data centers

4. The main goal of cloud computing is to ________, combine them to achieve higher throughput and be able to solve large scale computation problems.

A. reduces the overall cost

B. make a better use of distributed resources

C. include hundreds of thousands of computer servers

D. has super computing power

5. In cloud computing, users access the data, applications or any other services ________ regardless of the device used and the user's location.

A. with the help of a browser

B. with the large scale proliferation of the Internet

C. with powerful computing and data processing capacity

D. with minimal management effort

6. "Cloud computing" simply means "Internet computing", generally the Internet is seen as collection of clouds; thus the word cloud computing can be defined as ________ to the people and organizations.

A. eliminating the need to install and run the application

B. building the world's largest data centers

C. utilizing the Internet to provide technology enabled services

D. computing model for its major advantages

7. Google Apps is the paramount example of cloud computing, it enables users ________ and deployed on millions of machines over the Internet.

A. to access services via the browser

B. to provide technology enabled services

C. to install and run the application

D. to manage the operating systems

五、请将下述短文译成中文

1. The cloud computing is one kind of emerging business computing model. It will distribute the duty of calculation to the resource pool which the massive computers constitute, enables each kind of application system according to need to gain the computation strength, the storage space and all kinds of software service. This kind of resource pool is called "the cloud".

2. The cloud computing will concentrate all computation resources, and can be managed automatically through the software without intervene. The users who use cloud computing will not be worried by the tedious details and concentrate his own business. Cloud computing is a recent trend in IT that moves computing and data away from desktop and portable PCs into large data centers.

3. The major cloud providers such as Google, Microsoft and Amazon have built and are working on building the world's largest data centers across the United States and elsewhere. Each data center includes hundreds of thousands of computer servers, cooling equipment and substation power transformers.

4. Thousands of computers form a super server in cloud services which provide users with powerful computing and data processing capacity that is hard to realize for a personal computer. As a result this reduces the overall cost. In cloud computing, users access the data, applications or any other services with the help of a browser regardless of the device used and the user's location.

5. With the large scale proliferation of the Internet around the world, applications can now be delivered as services over the Internet. The main goal of cloud computing is to make a better use of distributed resources, combine them to achieve higher throughput and be able to solve large scale computation problems. Cloud computing has super computing power.

6. Cloud computing has been envisioned as the next-generation computing model for its major advantages in on-demand self-service, ubiquitous network access, location independent resource pooling and transference of risk. Cloud computing is the latest developments of computing models after distributed computing, parallel processing and grid computing.

7. "Cloud computing" simply means "Internet computing", generally the Internet is seen as collection of clouds; thus the word cloud computing can be defined as utilizing the Internet to provide technology enabled services to the people and organizations. Cloud computing enables consumers to access resources online through the Internet,

from anywhere at any time without worrying about technical/physical management and maintenance issues of the original resources.

8. A cloud application delivers "Software as a Service (SaaS)" over the Internet, thus eliminating the need to install and run the application on the users system. Softwares are provided as a service to the consumers according to their requirement, which enable consumers to use the services that are hosted on the cloud server. Google Apps is the most widely used SaaS.

9. The client need not purchase the required servers, data center or the network resources. Also the key advantage here is that customers need to pay only for the time duration they use the service. As a result customers can achieve a much faster service delivery with less cost.

参考译文

云计算

云计算是一种新出现的商务计算模式。它将计算任务分配给由大量计算机组成的资源池，并根据需要使每种应用系统获得计算能力、存储空间和各种软件服务。这种资源池被称为"云"。"云"是能够自我维护和自我管理的虚拟计算资源，通常是大规模的服务器群，包括计算服务器、存储服务器、宽带资源等。云计算集中所有计算资源并能够通过软件不间断地自动管理。使用云计算的用户不必担心乏味的细节并可以专注于自己的事情。云计算是信息技术的新趋势，它将计算及数据从桌面和手提电脑上转到大型数据中心。

像谷歌、微软和亚马逊这样大的云计算提供商在美国和世界其他地方已经建立了世界上最大的数据中心。每个数据中心包括成千上万个计算机服务器、冷却设备和变电站变压器。

随着互联网在全世界的大规模扩散，应用业务可以通过互联网来提供。云计算的主要目标是更好地利用和整合分布式资源以获得更高的数据吞吐量和解决大规模计算问题。云计算具有超强的计算能力。在云计算服务中，数千的计算机构成超级服务器，由这些服务器向用户提供强有力的数据计算和处理能力，这些能力是个人计算机难以实现的。这一结果是降低了整个费用。在云计算中，用户借助浏览器接入数据、应用或任何服务而不管所使用的设备和用户的位置。

云计算因其按需自我服务、无所不在的网络接入、与位置无关的资源池以及风险转移的主要优势而被看成是下一代计算模式。按需自我服务使用户自己能够通过使用云服务目录获得、配置和部署云服务而不需要IT的帮助。云计算是继分布式计算、并行计算和网格计算后最新发展的计算模式。由美国国家标准与技术研究院提供的云计算的定义：云计算是一种模型，它可以实现随时随地、便捷地、随需应变地从可配置计算资源共享池中获取所需的资源(例如，网络、服务器、存储应用及服务)，资源能够快速供应并释放，使管理资源的工作量和与服务提供商的交互减小到最低限度。

云计算就是互联网计算，通常互联网被看成是云的集合，这样云计算可以定义为使用互联网向个人和组织机构提供服务的技术。云计算使得顾客能够随时随地通过互联网在线接入资源，而不必担心最初资源的技术或物理的管理和维护。谷歌应用是最突出的云计算实例，它通过浏览器接入服务并部署在互联网中数百万的计算机上。云计算比其他计算模型更便宜，维护成本为零，因为服务提供商负责服务的有效使用而顾客不受资源设备管理和维护的影响。所部署的主要服务模型有软件即服务、平台即服务和基础设施即服务。

图 7-1(见第 89 页)表示了云计算结构中不同的分层。

云客户由依赖提供云计算服务的计算机硬件和软件组成。

云应用通过互联网提供软件即服务(SaaS)，这样用户就不必在其用户系统上安装和运行应用软件了。根据客户的需要，软件作为一种服务提供给客户，使用户能够使用托管在云服务器的服务。谷歌应用是最普遍使用的 SaaS。

平台服务"平台即服务(PaaS)"提供使用云结构的计算平台。向客户提供平台接入，使客户能将其定制的软件和其他应用放在云中。通过这种服务，开发者可以对所有系统和环境所需软件生命周期进行掌控，如软件的开放、测试、部署和网页应用的托管等。

基础设施服务"基础设施即服务(IaaS)"提供基础设施服务。可以出租计算处理、存储、网络容量和其他基本的计算资源，使客户能够管理操作系统、应用、存储和网络连接。客户不必购买所需要的服务器、数据中心或网络资源。其主要的优势在于客户只需对使用服务的这段时间付费。其结果是客户用很少的成本获得快得多的服务。

UNIT 8

PASSAGE

Edge Computing[1]

The cloud computing paradigm is a service provisioning model that provides user access to scalable distributed capabilities including computing, networking, and storage in the cloud data centers.[2] Cloud service providers (CSPs) provide flexibility and efficiency for end users by providing services such as software as a service (SaaS), platform as a service (PaaS) and infrastructure as a service (IaaS). For example, service vendors can scale services to fit their needs, customize applications and access cloud services from anywhere with an Internet connection. Thus cloud-based services are ideal for businesses with growing or fluctuating Internet bandwidth demands. Moreover, with cloud computing, enterprise users can ship applications to market more quickly, without worrying about underlying infrastructure costs, maintenance, disaster recovery, and automatic software updates.

To leverage the benefits of cloud computing, various deployment models including private cloud, public cloud, and hybrid cloud, are key factors for system reliability and scale for business needs.[3] Cloud computing has changed the way of business of all vertical domains as well as human being's daily life dramatically. Therefore, it's predicted that cloud service is inevitably becoming pervasive and ubiquitous in any commercial or personal market, which is similar to the prevalent dominance of Internet in nowadays.[4]

Although cloud computing can provide organizations dynamic, cloud-based operating models for cost optimization and increased competitiveness, it also has some disadvantages in many scenarios like industrial IoT, connected autonomous vehicles (CAVs), smart homes, and smart cities. For example, cloud computing based processing requires huge volume of data transportation from end devices and sensors, which consumes large network bandwidth.[5] Moreover, cloud data center based analysis is not possible for huge data generated from thousands of millions of end devices due to the incapability of computing and storage. Therefore, cloud computing based processing can't provide prompt responsiveness and short latency for big data analytics from massive IoT devices. In some scenarios where data privacy and security is the first concern, cloud computing data centers are not trustful to conduct the data analytics.[6]

Edge computing has emerged as a promising paradigm that provides capabilities of processing or storing critical data locally and pushing all received data to a central data center or cloud storage repository.[7] For example, in IoT use cases, the edge devices collect data from sensors and process it there, or send it back to a data center or the cloud for processing if the local processing power is not enough. To this end, edge computing paradigm can take some of the load off the central cloud data centers and migrate the tasks from cloud computing centers to network edge devices, reducing or even eliminating the processing workload at the central location.[8] The demand for scalable real-time data analytics in IoT scenarios is the main driving force for edge computing. In edge computing environment, data generation and consumption are concentrated to the edge of the network in many applications of smart home, smart city, and industrial Internet.

With the advances of IoT, 5G communication, autonomous driving, and smart cities, edge computing is connecting and bridging the gap between numerous end devices and the centralized cloud computing data centers. Moreover, in some cases where data privacy and security is the main concern, edge computing promises to provide data privacy preservation by keeping data inside the network edge rather than sending the data to centralized cloud data centers, which in turn provides lower latency, increases reliability and improves overall network efficiency.[9]

Since hundreds of millions of edge devices are geographically deployed in a distributed manner, and the data processing is also performed on heterogeneous distributed devices, it's very important to design new system architecture suitable for edge computing environments.[10] The data volume generated by various applications and devices running in the edge computing environment is huge and highly heterogeneous.[11] With the ever-increasing deployment of various IoT devices, lots of mobile devices and applications need more stringent requirements on service quality and real-time responsiveness of data processing.

Three popular application cases are listed as follows.

(1) IoT

The wide deployment of IoT devices and the increasing commercial demand for real-time data processing and the high quality of service of user experiences urge the creation of edge computing. Since more and more intelligent devices and sensors are deployed in the IoT environment, data production and consumption are performed and shifted to the edge of network gradually, which also needs elaborate computing technology for real-time analytics and pervasive processing.[12]

(2) SMART HOME

Technologies for better quality life and living conditions have changed human being's lifestyle and life quality. Deploying various sensors at home and sending collected data to remote cloud data center for processing introduce high risk of private data leakage, data abuse, and physical threat to massive local residents.[13] Therefore, the traditional cloud computing based data processing is not suitable for smart home applications, and data privacy preservation enabled edge computing emerges as the perfect alternative to smart home.

(3) SMART CITY

City is the place that consists of many smart homes, which implies that the edge computing paradigm can be extended from family level to city level, i.e., the smart city.[14] However, since a typical city also produces large volume of public services related data, even the most advanced cloud data centers can't process these data in real time for city-scale interactive analytics due to the lack of capabilities of computing, storage, and networking.[15] If data processing can be offloaded to the edge of the network, it can reduce the pressure of cloud data centers and make it possible for near real-time analytics. Moreover, in a smart city, one of the most important application scenarios is intelligent transportation. Networked traffic sensors and cameras provide perfect platform for edge data processing close to local data source, which makes it possible to solve traffic problems facing the urban residents, from traffic conditions alerts to road conditions prediction.[16]

Although the traditional cloud computing technology cannot meet requirements in terms of real-time response, privacy protection, and less energy consumption, the edge computing paradigm is not in essence replacing the cloud computing technology. In contrast, the cloud computing and edge computing are complementary and mutually reinforcing each other in many scenarios. Moreover, the edge computing and cloud computing will collaborate in the networked computing environment including scenarios such as IoT, smart city, smart home, industrial Internet, connected autonomous vehicles, etc.[17] The edge computing technology can fully exploit the computing capabilities of the edge devices performing partial or whole computing at the edge devices, thereby reducing the computing demand of the cloud data centers and the transmission bandwidth of core network. The collaboration of edge computing and cloud computing provides more opportunities for pervasive data analytics in IoT and low latency computing for latency critical applications such as autonomous driving and industrial networked systems.

Edge computing is emerging as one of the strategic technology that will redefine the future computing paradigm for its promise of lower latency, less bandwidth usage and data privacy protection.[18]

NEW WORDS AND PHRASES

edge	*n*. 边缘,刀刃
paradigm	*n*. 范例,范式
provision	*n*. 规定,条款,准备;*v*. 供给(食物及必需品)
scalable	*a*. 可扩展的,可攀登的

distribute	*vt*.分配,散布
flexibility	*n*.灵活性,柔韧性
platform	*n*.平台,站台
customize	*vt*.定制,量身定制
fluctuate	*v*.波动,起伏
disaster	*n*.灾难,灾害
automatic	*a*.自动的,机械的
leverage	*v*.利用;*n*.杠杆
deployment	*n*.部署,展开
hybrid	*a*.混合的,杂种的
reliability	*n*.可靠性,可靠度
vertical	*a*.垂直的,直立的
dramatically	*ad*.显著的,剧烈的
inevitable	*a*.必然的,不可避免的
pervasive	*a*.普遍的,到处渗透的
ubiquitous	*a*.普遍存在的,无所不在的
prevalent	*a*.普遍的,流行的
dominance	*n*.优势,统治
dynamic	*a*.动态的,动力的
optimization	*n*.最佳化,最优化
scenarios	*n*.场景,情节
consume	*v*.消费,消耗
incapability	*n*.无能力,无资格
latency	*n*.潜伏,延迟,时延
security	*n*.安全,安全性
repository	*n*.仓库,知识库
eliminate	*vt*.消除,排除
environment	*n*.环境,外界
bridge	*vt*.架桥,渡过
privacy	*n*.隐私,秘密
preservation	*n*.保存,保留
geographical	*a*.地理的,地理学的
heterogeneous	*a*.混杂的,异质的
stringent	*a*.严格的,严厉的
elaborate	*a*.精心制作的,详尽的

alternative a.选择性的,交替的
offload vt.卸载,分流
analytics n.解析学,分析论

NOTES

1. 题目是边缘计算。本文对云计算及边缘计算的基本概念做了说明。

2. that provides user access to scalable distributed capabilities 由 that 引导定语从句,修饰前面的 model;including computing, networking, and storage in the cloud data centers 现在分词短语,说明 capabilities。本句译文:云计算范例是一种业务提供模型,它为用户提供可扩展的分布式功能的访问,包括云数据中心的计算功能、联网功能和存储功能。

3. to leverage the benefits of cloud computing 不定式短语表示目的,为利用云计算的好处;including private cloud, public cloud, and hybrid cloud 现在分词短语做后置定语,修饰前面的 models。本句译文:为了利用云计算的优点,各种部署模型(包括私有云、公共云和混合云)是系统可靠性和业务需求可扩展性的关键因素。

4. it's … that 此句型中,it 是形式主语;which 引导非限定性定语从句。本句译文:因此,可以预见云服务在任何商业或个人市场中都将不可避免地变得普遍、无所不在,这与当今互联网的普遍主导地位类似。

5. which 引导非限定性定语从句。本句译文:例如,基于云计算的处理需要从终端设备和传感器传输大量数据,这会消耗大量网络带宽。

6. where data privacy and security is the first concern 是由 where 引导的定语从句,修饰 scenarios。本句译文:在某些应用场景下,数据隐私和安全是首要考虑的问题,云计算数据中心并不能得到信任进行数据分析。

7. that provides capabilities of processing or storing critical data locally and pushing all received data to a central data center or cloud storage repository 是由 that 引导的定语从句,修饰 paradigm。本句译文:边缘计算是一个很有前途的范例,它提供了本地处理或存储关键数据的能力,并将所有接收到的数据推送到一个中央数据中心或云存储库。

8. to this end 不定式短语表示目的;主句有两个谓语;reducing or even eliminating the processing workload at the central location 分词短语,表示伴随情况。本句译文:为了这个目的,边缘计算范例可以减轻中央云数据中心的部分负载,并将任务从云计算中心迁移到网络边缘设备,从而减少甚至消除中央位置的处理工作负载。

9. where data privacy and security is the main concern 是由 where 引导的定语从句,修饰 cases;介词短语 by 表示方式;which 引导非限定性定语从句。本句译文:此外,在数据隐私和安全是主要问题的情况中,边缘计算承诺提供数据隐私保护是通过将数据保持在网络边缘,而不是将数据发送到云数据中心,进而提供更低的延迟,增加可靠性,提高整体网络效率。

10. 由 since 引导原因状语从句。本句译文:由于数以百万计的边缘设备以分布的方式进行地理部署,并且数据处理也是在异构分布式设备上执行的,所以设计适合边缘计算

环境的新型系统结构就非常重要了。

11. generated by various applications and devices 过去分词短语做后置定语，修饰 volume；running in the edge computing environment 现在分词短语修饰 devices。本句译文：运行在边缘计算环境中的各种应用程序和设备产生的数据是大量和高度异构的。

12. Since more and more intelligent devices and sensors are deployed in the IoT environment 是由 since 引导的原因状语从句；本句是被动句；由 which 引导非限定性定语从句。本句译文：随着越来越多的智能设备和传感器部署在物联网环境中，数据生产和消费逐渐进行并转移到网络边缘，这也需要精细的计算技术来进行实时分析和广泛处理。

13. Deploying various sensors at home and sending collected data to remote cloud data center 并列的两个动名词短语做主语。本句译文：在家中部署各种传感器并将收集到的数据发送到远程云数据中心进行处理，会给大量本地居民带来私人数据泄露、数据滥用和物理威胁的高风险。

14. that consists of many smart homes 是由 that 引导的定语从句，修饰 place；由 which 引导非限定性定语从句；that the edge computing paradigm can be extended from family level to city level, i. e. , the smart city 是由 that 引导的宾语从句，做动词 implies 的宾语。本句译文：城市是由许多智能家居组成的地方，这意味着边缘计算范例可以从家庭层面扩展到城市层面，即智慧城市。

15. 由 since 引导原因状语从句；due to，由于。本句译文：然而，由于一个典型的城市也会产生大量的公共服务相关数据，即使是最先进的云数据中心也无法实时处理这些数据，以进行城市规模的交互式分析，这是由于缺乏计算、存储和联网的能力。

16. close to，靠近，接近；which 引导非限定性定语从句；facing the urban residents 现在分词短语做后置定语，修饰 problems。本句译文：网络化的交通传感器和摄像头为接近本地数据源的边缘数据处理提供了完美的平台，使解决城市居民面临的交通问题成为可能，从交通状况预警到道路状况预测。

17. including scenarios 现在分词短语做后置定语，修饰前面的 environment；such as，比如。本句译文：此外，边缘计算和云计算将在联网计算环境中进行协作，包括物联网、智能城市、智能家居、工业互联网、联网自动驾驶汽车等场景。

18. that will redefine the future computing paradigm 是由 that 引导的定语从句，修饰前面的 technology。本句译文：边缘计算正在成为一种战略性技术，它因为有希望带来更低的延迟、更少的带宽使用和数据隐私保护，将会重新定义未来的计算范式。

EXERCISES

一、请将下述词组译成英文

1. 私人数据泄露的高风险　2. 智能家居应用　3. 数据隐私保护
4. 提供快速响应　5. 云数据计算中心　6. 系统可靠性的关键因素
7. 数据隐私和保护　8. 从传感器收集数据　9. 在边缘计算环境中
10. 消除处理负担　11. 提供低延时　12. 各种 IoT 设备的部署
13. 催生边缘计算　14. 海量的数据传输　15. 边缘计算

16. 云计算范例　　17. 业务提供模式　　18. 可扩展的分布能力

二、请将下述词组译成中文

1. the incapability of computing and storage
2. the perfect alternative to smart home
3. the most advanced cloud data centers
4. be offloaded to the edge of the network
5. to reduce the pressure of cloud data centers
6. the most important application scenarios
7. to leverage the benefits of cloud computing
8. the prevalent dominance of Internet in nowadays
9. the data volume generated by various applications
10. to send collected data to remote cloud data center
11. to provide perfect platform for edge data processing
12. to customize applications and access cloud services
13. the cost optimization and increased competitiveness
14. to produce large volume of public services related data
15. to provide capabilities of processing or storing critical data
16. the increasing commercial demand for real-time data processing
17. the stringent requirements on service quality and real-time responsiveness

三、选择合适的答案填空

1. Cloud service providers (CSPs) provide flexibility and efficiency for end users by ________ services such as software as a service (SaaS), platform as a service (PaaS) and infrastructure as a service (IaaS).

A. to provide　　B. providing
C. provided　　D. provide

2. Thus cloud-based services are ideal for businesses with growing or ________ Internet bandwidth demands.

A. fluctuate　　B. to fluctuate
C. fluctuated　　D. fluctuating

3. To this end, edge computing paradigm can take some of the load off the central cloud data centers and migrate the tasks from cloud computing centers to network edge devices, ________ or even eliminating the processing workload at the central location.

A. reducing　　B. reduce
C. to reduce　　D. reduced

4. Since hundreds of millions of edge devices are geographically deployed in a distributed manner, and the processing to data is also performed on heterogeneous distributed devices, it's very important ________ new system architecture suitable for edge computing environments.

A. design　　B. designing

C. to design　　D. designed

5. Networked traffic sensors and cameras provide perfect platform for edge data processing close to local data source, which makes it possible to solve traffic problems ________ the urban residents, from traffic conditions alerts to road conditions prediction.

A. to face　　B. faced

C. face　　D. facing

6. Moreover, the edge computing and cloud computing will collaborate in the networked computing environment ________ scenarios such as IoT, smart city, smart home, industrial internet, connected autonomous vehicles, etc.

A. including　　B. to include

C. included　　D. include

7. The edge computing technology can fully exploit the computing capabilities of the edge devices performing partial or whole computing at the edge devices, thereby ________ the computing demand of the cloud data centers and the transmission bandwidth of core network.

A. reduced　　B. reducing

C. to reduce　　D. reduce

四、根据课文内容选择答案

1. The cloud computing paradigm is a service provisioning model that provides user ________ including computing, networking, and storage in the cloud data centers.

A. providing services such as software as a service

B. access to scalable distributed capabilities

C. deployment models including private cloud

D. optimization and increased competitiveness

2. Cloud computing has changed the way of ________ as well as human being's daily life dramatically.

A. benefits of cloud computing

B. key factors for system reliability

C. storage in the cloud data

D. business of all vertical domains

3. Although cloud computing can provide organizations dynamic, cloud-based operating models for ________, it also has some disadvantages in many scenarios like industrial IoT, connected autonomous vehicles (CAVs), smart homes, and smart cities.

A. cost optimization and increased competitiveness

B. huge volume of data transportation

C. cloud computing data centers

D. data generation and consumption

4. Therefore, cloud computing based processing can't ________ for big data analytics from massive IoT devices.

A. real-time data analytics in IoT scenarios

B. capabilities of processing or storing critical data l

C. provide prompt responsiveness and short latency

D. the centralized cloud computing data centers

5. For example, in IoT use cases, the edge devices collect data from sensors and process it there, or send it back to a data center or the cloud for processing if ________.

A. the local processing power is not enough

B. edge computing is connecting and bridging the gap

C. edge devices are geographically deployed

D. data privacy and security is the main concern

6. With the advances of IoT, 5G communication, autonomous driving, and smart cities, edge computing is connecting and bridging the gap between ________.

A. increase reliability and improves overall network efficiency

B. huge volume of data transportation from end devices and sensors

C. operating models for cost optimization and increased competitiveness

D. numerous end devices and the centralized cloud computing data centers

7. With the ever-increasing deployment of various IoT devices, lots of mobile devices and applications need more stringent requirements on ________ of data processing.

A. prompt responsiveness and short latency

B. incapability of computing and storage

C. service quality and real-time responsiveness

D. end devices and sensors

8. Although the traditional cloud computing technology cannot meet requirements in terms of real-time response, privacy protection, and less energy consumption, the edge computing paradigm is not in essence ________.

A. operating models for cost optimization

B. replacing the cloud computing technology

C. eliminating the processing workload

D. connecting and bridging the gap

五、请将下列短文译成中文

1. Cloud service providers (CSPs) provide flexibility and efficiency for end users by providing services such as software as a service (SaaS), platform as a service (PaaS) and infrastructure as a service (IaaS). For example, service vendors can scale services to

fit their needs, customize applications and access cloud services from anywhere with an Internet connection. Thus cloud-based services are ideal for businesses with growing or fluctuating Internet bandwidth demands.

2. Although cloud computing can provide organizations dynamic, cloud-based operating models for cost optimization and increased competitiveness, it also has some disadvantages in many scenarios like industrial IoT, connected autonomous vehicles (CAVs), smart homes, and smart cities. For example, cloud computing based processing requires huge volume of data transportation from end devices and sensors, which consumes large network bandwidth.

3. With the advances of IoT, 5G communication, autonomous driving, and smart cities, edge computing is connecting and bridging the gap between numerous end devices and the centralized cloud computing data centers. Moreover, in some cases where data privacy and security is the main concern, edge computing promises to provide data privacy preservation by keeping data inside the network edge rather than sending the data to centralized cloud data centers, which in turn provides lower latency, increases reliability and improves overall network efficiency.

4. Since hundreds of millions of edge devices are geographically deployed in a distributed manner, and the processing to data is also performed on heterogeneous distributed devices, it's very important to design new system architecture suitable for edge computing environments. The data volume generated by various applications and devices running in the edge computing environment is huge and highly heterogeneous.

5. Deploying various sensors at home and sending collected data to remote cloud data center for processing introduce high risk of private data leakage, data abuse, and physical threat to massive local residents. Therefore, the traditional cloud computing based data processing is not suitable for smart home applications, and data privacy preservation enabled edge computing emerges as the perfect alternative to smart home.

6. However, since a typical city also produces large volume of public services related data, even the most advanced cloud data centers can't process these data in real time for city-scale interactive analytics due to the lack of capabilities of computing, storage, and networking. If data processing can be offloaded to the edge of the network, it can reduce the pressure of cloud data centers and make it possible for near real-time analytics.

7. Although the traditional cloud computing technology cannot meet requirements in terms of real-time response, privacy protection, and less energy consumption, the edge computing paradigm is not in essence replacing the cloud computing technology. In contrast, the cloud computing and edge computing are complementary and mutually reinforcing each other in many scenarios.

8. The edge computing technology can fully exploit the computing capabilities of

the edge devices performing partial or whole computing at the edge devices, thereby reducing the computing demand of the cloud data centers and the transmission bandwidth of core network. The collaboration of edge computing and cloud computing provides more opportunities for pervasive data analytics in IoT and low latency computing for latency critical applications such as autonomous driving and industrial networked systems.

参考译文

边缘计算

云计算范例是一种业务提供模型,它为用户提供可扩展的分布式功能的访问,包括云数据中心的计算功能、联网功能和存储功能。云服务提供商(CSPs)通过提供软件即服务(SaaS)、平台即服务(PaaS)和基础设施即服务(IaaS)等服务,为最终用户提供业务灵活性和高效率。例如,服务供应商可以扩展服务来满足他们的需求,定制应用程序,并在任何有互联网连接的地方访问云服务。因此,对于互联网带宽需求有增长或变动的企业来说,基于云的服务是理想的选择。此外,通过云计算,企业用户可以更快地将应用服务推向市场,而无须担心底层基础设施成本、维护、故障恢复和自动软件更新。

为了利用云计算的优点,各种部署模型(包括私有云、公共云和混合云)是系统可靠性和业务需求可扩展性的关键因素。云计算已经极大地改变了所有垂直领域的业务方式以及人类的日常生活。因此,可以预见云服务在任何商业或个人市场中都将不可避免地变得普遍、无所不在,这与当今互联网的普遍主导地位类似。

尽管云计算可以为企业提供动态的、基于云的运营模型,以实现成本优化和提高竞争力,但在工业物联网、联网自动车辆(CAVs)、智能家居和智能城市等许多场景中,它也有一些缺点。例如,基于云计算的处理需要从终端设备和传感器传输大量数据,这会消耗大量网络带宽。此外,基于云数据中心的分析,由于计算和存储能力的不足,不可能对数以百万计的终端设备产生的巨大数据进行分析。因此,基于云计算的处理无法为来自大量物联网设备的大数据分析提供快速响应和短延迟。在某些应用场景下,数据隐私和安全是首要考虑的问题,云计算数据中心并不能得到信任进行数据分析。

边缘计算是一个很有前途的范例,它提供了本地处理或存储关键数据的能力,并将所有接收到的数据推送到一个中央数据中心或云存储库。例如,在物联网应用中,边缘设备从传感器收集数据并在那里进行处理,如果本地处理能力不足,则将数据发送回数据中心或云进行处理。为此,边缘计算范型可以减轻中央云数据中心的部分负载,并将任务从云计算中心迁移到网络边缘设备,从而减少甚至消除中央位置的处理工作负载。物联网场景中对可扩展实时数据分析的需求是边缘计算的主要驱动力。在边缘计算环境中,智能家居、智慧城市、工业互联网等诸多应用中,数据的生成和消费都集中在网络的边缘。

随着物联网、5G通信、自动驾驶和智慧城市的发展,边缘计算正在连接和弥合众多终端设备和集中式云计算数据中心之间的差距。此外,在数据隐私和安全是主要问题的情况下,边缘计算承诺提供数据隐私保护是通过将数据保持在网络边缘,而不是将数据发送到云数据中心来实现的，进而提供更低的延迟，增加可靠性，提高整体网络效率。

由于数以百万计的边缘设备以分布的方式进行地理部署，以及处理数据也是在异构分布式设备上执行的，所以设计适合边缘计算环境的新型系统结构非常重要。由运行在边缘计算环境中的各种应用程序和设备产生的数据是大量和高度异构的。随着各种物联网设备的不断部署，大量移动设备和应用对服务质量和数据处理的实时响应能力提出了更高的要求。

以下列举了三个流行的应用案例：物联网、智慧家庭和智能城市。

1. 物联网

物联网设备的广泛部署、日益增长的对实时数据处理的商业需求以及用户高质量体验的需求，催生了边缘计算。随着越来越多的智能设备和传感器部署在物联网环境中，数据生产和消费逐渐执行并转移到网络的边缘，这也需要精细的计算技术来进行实时分析和广泛处理。

2. 智慧家庭

改善生活质量和生活条件的技术已经改变了人类的生活方式和生活质量。在家中部署各种传感器并将收集到的数据发送到远程云数据中心进行处理，会给大量本地居民带来私人数据泄露、数据滥用和物理威胁的高风险。因此，传统的基于云计算的数据处理不适合于智能家居应用，数据隐私保护的边缘计算成为智能家居的完美替代品。

3. 智能城市

城市是由许多智能家居组成的地方，这意味着边缘计算范例可以从家庭层面扩展到城市层面，即智慧城市。然而，由于一个典型的城市也会产生大量的公共服务相关数据，即使是最先进的云数据中心也无法实时处理这些数据，以进行城市规模的交互式分析，这是由于缺乏计算、存储和联网的能力。如果数据处理可以被转移到网络的边缘，它可以减少云数据中心的压力，使近乎实时的分析成为可能。此外，在智慧城市中，智能交通是最重要的应用场景之一。网络化的交通传感器和摄像头为接近本地数据源的边缘数据处理提供了完美的平台，从交通状况预警到道路状况预测，使解决城市居民面临的交通问题成为可能。

虽然传统的云计算技术在实时响应、隐私保护、低能耗等方面无法满足需求，但边缘计算范式在本质上并不是要取代云计算技术。相比之下，云计算和边缘计算在许多场景中是互补的、相辅相成的。此外，边缘计算和云计算将在联网计算环境中进行协作，包括物联网、智能城市、智能家居、工业互联网、联网自动驾驶汽车等场景。边缘计算技术可以充分利用边缘设备的计算能力，在边缘设备上进行部分或整体计算，从而降低云数据中心的计算需求和核心网络的传输带宽。边缘计算和云计算的合作为物联网无所不在的数据分析和延迟关键应用（如自动驾驶和工业网络系统）的低延迟计算提供了更多的机会。

边缘计算正在成为一种战略性技术，它因为有希望带来更低的延迟、更少的带宽使用和数据隐私保护，将会重新定义未来的计算范式。

UNIT 9

PASSAGE

The Development and Application of IoT[1]

The car will automatically alert when the driver's operation error; the briefcase will remind the owner what they forgot to take; clothes will "tell" the washing machine on the color and water temperature requirements and so on. Those are the vision of the IoT (internet of things) era which is described by the ITU (International Telecommunications Union).[2] Our society has a lot of applications of things-things connected, such as the transport vehicles which equipped with GPS and RFID chip, car through the roll station with no stop, then automatically unloaded with no people.

IoT was first introduced in 1999 and was initially designed as a mechanism that could link the virtual world with the physical world.[3] However, it has significantly evolved since its early manifestation and now incorporates many different components across numerous industries and applications to the extent that it has become a significant part of everyday life in contemporary society.

The concept of IoT is regarded as another scientific and economic tide in the global information industry after the internet, having attracted highly attention of governments, enterprises, and academia.[4] America, Europe and Japan even enlist it in their national and informational strategy. The IoT is a kind of expanding and extending of network technology, with the basis of internet technology, whose core and basis are internet technology. The definition of IoT refers to accordance with the appointed agreement having any objects connected with Internet to exchange information and communicate to realize intelligent recognition, orientation tracing, monitor and management through the information peripheral equipments, such as RFID, infrared sensor, GPS and laser scanner etc.

The key technologies of IoT include radio frequency identification (RFID) devices, WSN networks, infrared sensors, global positioning systems, internet and mobile network. As the most important technology of the IoT, RFID technology plays a very important role in the development of the IoT. RFID is still faced with many problems to be solved.

The "things" in the IoT not only should be achieved "connected together", but also can be realized the functions of recognition, localization, tracing, management and so on, which requires all things must be identified.[5] The RFID technology is the "speaking technology" for these things, therefore the RFID technology possessed an outstanding status in these key technologies. Radio frequency identification (RFID) technology is one kind of automatic diagnosis technology emerged in 1990s, it can realize non-contact information transmission and recognition by radio-frequency signal through the space coupling. A typical RFID system consists of RFID tag, reader and the application systems. The tag is mainly composed of antenna, resonant capacitance and IC chip; the reader usually contains radio-frequency signal launch unit, radio frequency receiving element and control unit.

RFID system consists of data acquisition and back-end database network applications. Already published or under development standards are mainly related to data collection including the interface of electronic tag and reader, the data exchange protocol between the reader and computers, the performance of RFID tag and reader, the content of RFID tag data encoding standards.[6]

Wireless sensor network (WSN) is made up by the large number of low-cost micro sensor nodes deployed in the monitoring area, and form multi-hop ad hoc networks with the wireless communication.[7] Sensor network will be able to expand the ability of people to interact with the real world remotely. Wireless sensor network is a new platform for information access.

At the background of the integration of industrialization and informationization promoted by the government, the IoT will be the realistic start point of industry and other information industry. Once the massive popularity of IoT, many more small items need to install intelligent sensors. The number of the sensors and electronic tags which used for animals, plants, machinery and other items will greatly exceed the current phone number. According to the current demand on the Internet of Things, there are about billions of sensors and electronic tags in recent years. Experts predict that by 2020, embedded chips, sensors, radio frequency and other "smart objects" will be over 1 trillion.[8]

While most hospitals today are full of smart devices, few of the sensors in these devices communicate with each other. Once these sensors are fully connected via the IoT, the practice of healthcare will be dramatically transformed. For example, the IoT will be able to warn patients—at home via Amazon's Alexa or on a smartphone—of blood clots before impending strokes or heart attacks. Sensors linked to electronic medical records will allow the IoT to quickly diagnose a patient's likely physical state to assist emergency medical personnel and expedite treatment.[9]

Another core industry that is likely to be radically transformed is the transportation industry.[10] As short as five years ago, most of us would have thought the notion of driverless cars was either science fiction, or at a minimum, decades away. Yet, today driverless cars are a reality as Google is currently testing these autonomous vehicles on open roads and traditional automakers such as Mercedes-Benz, Nissan, and Audi are revamping their business strategies around this transportation game-changer. Because each driverless car is a collection of smart sensor devices that can be interconnected into a holistic network, the IoT will be able to aggregate and leverage the collective intelligence distributed throughout the network to drive all the vehicles concurrently.[11]

The term industrial Internet of things (IIoT) is often encountered in the manufacturing industries, referring to the industrial subset of the IoT. IIoT in manufacturing could generate so much business value that it will eventually lead to the fourth industrial revolution, so the so-called Industry 4.0.[12] It is estimated that in the future, successful companies will be able to increase their revenue through Internet of things by creating new business models and improve productivity. The potential of growth by implementing IIoT may generate $ 12 trillion of global GDP by 2030.

IoT devices are a part of the larger concept of home automation, which can include lighting, heating and air conditioning, media and security systems.[13] Long term benefits could include energy savings by automatically ensuring lights and electronics are turned off.[14]

A smart home or automated home could be based on a platform or hubs that control smart devices and appliances.[15] For instance, using Apple's HomeKit, manufacturers can have their home products and accessories controlled by an application in iOS devices such as the iPhone and the Apple Watch. This could be a dedicated App or iOS native applications such as Siri. There are also dedicated smart home hubs that are offered as standalone platforms to connect different smart home products and these include the Amazon Echo, Google Home, Apple's HomePod, and Samsung's SmartThings Hub.[16]

Another key application of smart home is to provide assistance for those with disabilities and elderly individuals. These home systems use assistive technology to accommodate an owner's specific disabilities. Voice control can assist users with sight and mobility limitations while alert systems can be connected directly to cochlear implants worn by hearing impaired users.[17] They can also be equipped with additional safety features. These features can include sensors that monitor for medical emergencies such as falls or seizures.[18] Smart home technology applied in this way can provide users with more freedom and a higher quality of life.

NEW WORDS AND PHRASES

automatically　*ad*. 自动地
alert　*v*. 向……报警
briefcase　*n*. 公文包，手提包
temperature　*n*. 温度
vision　*n*. 视力，美景
describe　*vt*. 描述

vehicle	*n*.车辆,交通工具
unload	*v*.卸下,卸货
initially	*ad*.最初地,首先地
mechanism	*n*.机制,原理,结构
virtual	*a*.虚拟的
manifestation	*n*.表示,显示,示威
incorporate	*v*.包含,吸收,合并
numerous	*a*.为数众多的,许多的
contemporary	*a*.同时代的,当代的
tide	*n*.潮流,趋势
attract	*v*.吸引,诱惑
enterprise	*n*.企业,事业
academia	*n*.学术界,学术生活
enlist	*v*.征募,使……入伍
expand	*v*.扩张,张开,详述
strategy	*n*.战略
core	*n*.核心
definition	*n*.定义
refer to	涉及,指的是
object	*n*.物体,目标,对象
intelligent	*a*.智能的
orientation trace	定向跟踪
peripheral	*a*.(*n*.)外围的,次要的;外围设备
infrared	*a*.(*n*.)红外线的;红外线
sensor	*n*.传感器
scanner	*n*.扫描器,扫描设备
identification	*n*.鉴定,识别,身份证明
achieve	*v*.获取,获得
not only but also	不仅……而且
recognition	*n*.识别,承认,认出
localization	*n*.定位,地方化
identify	*v*.确定,鉴定,识别
possess	*vt*.拥有,持有,支配
outstanding	*a*.杰出的,突出的
status	*n*.地位,情形,状况

diagnosis	n.诊断,分析
non-contact	n.非接触
couple	v.结合,连接,耦合
tag	n.标签,名称
resonant	a.洪亮的,回响的
antenna	n.天线
capacitance	n.电容
acquisition	n.获得,购置物
back-end	n.后端后台
performance	n.性能,表演,执行
include	vt.包括,包含
content	n.内容,目录
micro	a.微小的,基本的
deploy	v.部署,展开
hop	v.单足跳行

NOTES

1. 本文内容涉及物联网 IoT。题目译成:物联网 IoT 的发展与应用。

2. which 引导定语从句,从形式上来看是限定性定语从句,句子也比较短,按照和译法翻译较好,译成:……的。本句译文:这就是 ITU 所描述的物联网 IoT 的前景。

3. 此句有两个谓语,被动句可以按主动加来翻译;that 引导定语从句。本句译文:物联网开始于 1999 年,其开始的设想是要连接虚拟世界与物理世界。

4. 本句是被动句;分词短语 having attracted highly attention of governments, enterprises, and academia 表示一种伴随情况,也是对句子前面部分的说明。本句译文:物联网被看作是继互联网之后全球信息业的另一个科技浪潮,吸引了政府、企业和学术界的高度关注。

5. 本句是被动句;not only…but also,不仅,而且;which requires all things must be identified 是由 which 引导非限定性定于从句,可以按照分译法来翻译。本句译文:在 IoT 中的"物"不仅要实现连在一起,也要实现识别、定位、追踪、管理等功能,这就要求能识别所有的物品。

6. 介词短语 including 后面的内容比较多,有四个并列的部分。此分词短语对句子前面部分做说明。本句译文:已经发布或正在开发的标准主要涉及数据采集,包括电子标签和阅读器接口、阅读器和计算机之间的数据交换协议、RFID 标签和阅读器的性能、RFID 标签数据编码标准内容。

7. 此句主语有两个谓语;过去分词短语 deployed in the monitoring area 做后置定语,修饰前面的 nodes;介词短语 with the wireless communication 做后置定语,修饰 networks。本句译文:无线传感器网络 WSN 是由大量廉价的部署在监控区域的微型传感器节点组成,并构成无线通信的多跳特设网络。

8. 句中 that 引导宾语从句,做 predict 的宾语。本句译文:专家估计到 2020 年,嵌入式芯片、传感器、无线射频和其他“智能物体”的数量将会超过 1 万亿。

9. linked to electronic medical records 过去分词短语做后置定语,修饰前面的 sensors。本句译文:连接到电子医疗记录仪的传感器允许 IoT 快速诊断病人的身体状态,以便帮助急诊医护人员并加快治疗。

10. that 引导定语从句,修饰前面的 industry;is likely to,有可能;to be radically transformed,不定式做表语,此处用被动形式。本句译文:另一有可能被彻底改变的行业是交通运输业。

11. because 引导原因状语从句,that 引导定语从句;to drive all the vehicles concurrently 不定式短语,表示目的。本句译文:因为每一辆无人驾驶汽车都是一个智能传感器设备的集合,这些设备彼此连接成为一个整体网络,IoT 能够聚集并利用分布在整个网络中的集体智能来驾驶汽车。

12. so…that,如此的,以至于;so-called 所谓的,号称的。本句译文:制造业的 IIoT 可能产生非常多的商业价值,这些最终会导致第四次工业革命,也就是所谓的工业 4.0。

13. which 引导非限定性定语从句。本句译文:IoT 设备是家庭自动化的一个组成部分,包括照明设备、取暖和空调设备、媒体和安全系统。

14. by 引导的介词短语表示手段,ensuring 做介词 by 的宾语,其后又跟随省略 that 的宾语从句。本句译文:长远来看,通过自动关闭照明和电子设备可以带来节能的好处。

15. that 引导定语从句,修饰前面的 hubs。本句译文:智能家庭或自动家庭可以以控制智能设备的平台或集线器为基础。

16. 此句中由 and 连接两个并列句;that 引导定语从句;to connect 不定式短语表示目的。本句译文:也有专用的智能集线器是作为独立平台来连接不同的智能家居产品,这些包括 Amazon Echo、Google Home、苹果公司的 HomePod 和三星公司的 SmartThings Hub。

17. with sight and mobility limitations 介词短语做后置定语,修饰前面的 users;worn 做后置定语修饰前面的词。本句译文:声音控制可以帮助那些视力受限和移动受限的人士,而告警系统可以直接连到听力障碍人士佩戴的人工耳蜗。

18. that 引导定语从句,修饰前面的 sensors;such as,例如。本句译文:这些特性包括监控紧急情况如跌倒和癫痫的传感器。

EXERCISES

一、请将下述词组译成英文

1. IoT 前景
2. GPS 和激光扫描仪
3. 空间耦合
4. 定向追踪
5. 声音控制
6. 红外线传感器
7. 报警系统
8. 听觉缺陷用户
9. 智能传感器设备
10. IoT 概念
11. 全球定位系统
12. 识别功能
13. 通过 IoT 连接
14. 无人驾驶汽车
15. 无线传感器网络
16. 视力和移动性限制
17. 附加的安全特性
18. 无人驾驶汽车的概念

19. 总的智能产品消费　20. 家庭自动化概念

二、请将下述词组译成中文

1. link the virtual world with the physical world
2. incorporate many different components
3. numerous industries and application
4. another scientific and economic tide
5. the global information industry
6. the national and informational strategy
7. to realize intelligent recognition
8. the information peripheral equipment
9. the radio-frequency identification devices
10. one kind of automatic diagnosis technology
11. to realize noncontact information transmission
12. the radio-frequency signal launch unit
13. the data acquisition and backend database
14. the interface of electronic tag and reader
15. the large number of low-cost sensor nodes
16. to interact with the real world
17. the embedded chips, sensors and other smart objects
18. to aggregate and leverage the collective intelligence
19. the term of industrial Internet of things
20. the fourth industrial revolution
21. to control smart devices and appliances
22. to connect different smart home products
23. the high adaptability to new products

三、选择合适的答案填空

1. IoT was first introduced in 1999 and was initially ________ as a mechanism that could link the virtual world with the physical world.

A. designed　　B. design

C. to design　　D. designing

2. The concept of IoT is regarded as another scientific and economic tide in the global information industry after the Internet, ________ attracted highly attention of governments, enterprises, and academia.

A. have　　B. to have

C. had　　D. having

3. The IoT is a kind of ________ and extending of network technology, with the basis of Internet technology, whose core and basis are Internet technology.

A. expand　　B. expanded

C. expanding　　D. to expend

4. The tag is mainly ________ of antenna, resonant capacitance and IC chip; the reader usually contains radio-frequency signal launch unit, radio frequency receiving element and control unit.

A. compose　　B. composed

C. to compose　　D. composing

5. Already published or under development standards are mainly related to data collection ________ the interface of electronic tag and reader , the data exchange protocol between the reader and computers , the performance of RFID tag and reader , content of RFID tag data encoding standards.

A. include　　B. to include

C. including　　D. included

6. Sensor network will be able ________ the ability of people to interact with the real world remotely. Wireless sensor network is a new platform for information access.

A. to expand　　B. expand

C. expanded　　D. expanding

四、根据课文内容选择答案

1. Our society has a lot of ________, such as the transport vehicles which equipped with GPS and RFID chip, car through the roll station with no stop, then automatically unloaded with no people.

A. washing machine on the color

B. applications of things-things connected

C. another scientific and economic tide

D. national and informational strategy

2. However, it has significantly evolved since its early manifestation and now ________ across numerous industries and applications to the extent that it has become a significant part of everyday life in contemporary society.

A. refers to accordance with the appointed agreement

B. link the virtual world with the physical world

C. equip with GPS and RFID chip

D. incorporates many different components

3. ________ include radio frequency identification (RFID) devices, WSN networks, infrared sensors, global positioning systems, internet and mobile network.

A. The key technologies of IoT

B. the basis of internet technology

C. The "things" in the IoT

D. one kind of automatic diagnosis technology

4. The tag is mainly composed of antenna, resonant capacitance and IC chip; the

reader usually contains ________, radio frequency receiving element and control unit.

A. non-contact information transmission and recognition

B. the interface of electronic tag and reader

C. the fourth industrial revolution

D. radio-frequency signal launch unit

5. IIoT in manufacturing could generate so much business value that it will eventually lead to ________, so the so-called Industry 4.0.

A. the practice of healthcare

B. a collection of smart sensor devices

C. the fourth industrial revolution

D. the notion of driverless cars

6. These home systems use assistive technology to ________.

A. include lighting, heating and air conditioning

B. accommodate an owner's specific disabilities

C. have their home products and accessories

D. be able to increase their revenue

7. These features can include sensors that monitor for medical emergencies such as falls or seizures. Smart home technology applied in this way can provide users with more freedom and ________.

A. a platform or hubs

B. standalone platforms to connect

C. a higher quality of life

D. a part of the larger concept

五、请将下列短文译成中文

1. The car will automatically alert when the driver's operation error; the briefcase will remind the owner what they forgot to take; clothes will "tell" the washing machine on the color and water temperature requirements and so on.

2. IoT was first introduced in 1999 and was initially designed as a mechanism that could link the virtual world with the physical world. However, it has significantly evolved since its early manifestation and now incorporates many different components across numerous industries and applications to the extent that it has become a significant part of everyday life in contemporary society.

3. The concept of IoT is regarded as another scientific and economic tide in the global information industry after the internet, having attracted highly attention of governments, enterprises, and academia. America, Europe and Japan even enlist it in their national and informational strategy.

4. The definition of IoT refers to accordance with the appointed agreement having any objects connected with Internet to exchange information and communicate to realize

intelligent recognition, orientation tracing, monitor and management through the information peripheral equipments, such as, RFID, infrared sensor, GPS and laser scanner etc.

5. The "things" in the IoT not only should be achieved "connected together", but also can be realized the functions of recognition, localization, tracing, management and so on, which requires all things must can be identified. The RFID technology is the "speaking technology" for these things, therefore the RFID technology possessed an outstanding status in these key technologies.

6. RFID system consists of data acquisition and back-end database network applications. Already published or under development standards are mainly related to data collection including the interface of electronic tag and reader, the data exchange protocol between the reader and computers, the performance of RFID tag and reader, content of RFID tag data encoding standards.

7. At the background of the integration of industrialization and informationization promoted by the government, the IoT will be the realistic start point of industry and other information industry. Once the massive popularity of IoT, many more small items need to install intelligent sensors. The number of the sensors and electronic tags which used for animals, plants, machinery and other items will greatly exceed the current phone number.

8. Another core industry that is likely to be radically transformed is the transportation industry. As short as five years ago, most of us would have thought the notion of driverless cars was either science fiction, or at a minimum, decades away.

参考译文

物联网的发展与应用

当司机驾驶汽车操作有误时,汽车会自动告警;公文包会提醒物主忘了拿什么东西;衣服会告诉洗衣机衣服的颜色及需要的水温等。这就是 ITU 所描述的物联网 IoT 的前景。我们的社会中有大量的物物连接的应用,如安装了 GPS 和 RFID 芯片的运输车辆,当没有乘客下车时就可以自动的过站不停。

物联网开始于 1999 年,其最初的设想是要连接虚拟世界与物理世界。然而,从早期一出现其发展就引人注目,现在很多不同行业都有其应用,它已经成为当下社会日常生活中非常重要的一部分。

物联网被看作是继互联网之后全球信息业的另一个科技和经济浪潮,吸引了政府、企业和学术界的高度关注。美国、欧洲和日本甚至将其列为国家信息战略。IoT 是以互联网为基础的网络技术的扩展和延伸,其核心和基础是互联网技术。物联网的定义是指:通过射频识别 RFID、红外感应器、全球定位系统 GPS、激光扫描器等信息传感设备,按照约定的协议,把任何物品与互联网连接,进行信息交换和通信,以实现智能化识别、定向追踪、监控和管理。

物联网的关键技术包括射频识别 RFID 设备、无线传感网络 WSN、红外传感器、全球定位系统、互联网和移动网络。射频识别 RFID 技术作为物联网技术中最重要的技术,在物联网的发展中发挥着重要的作用。RFID 仍然还面临许多问题要解决。

在 IoT 中的"物"不仅要实现连在一起,也要实现识别、定位、追踪、管理等功能,这就要求能识别所有的物品。射频识别 RFID 技术就是所有物品的"说话技术",因此 RFID 在这些关键技术中具有突出的重要地位。出现在 1990 年的 RFID 技术是一种自动诊断技术,它可以通过射频信号的空间耦合实现非接触的信息传输和识别。典型的 RFID 系统包括 RFID 标签、阅读器和应用系统。标签主要由天线、回响电容和 IC 芯片组成;阅读器通常包括射频信号发送单元、射频接收单元和控制单元。

RFID 系统由数据获取和后端数据库网络应用组成。已经发布或正在开发的标准主要涉及数据采集,包括电子标签和阅读器接口、阅读器和计算机之间的数据交换协议、RFID 标签和阅读器的性能、RFID 标签数据编码标准内容。

无线传感器网络 WSN 是由大量廉价的部署在监控区域的微型传感器节点组成,并构成无线通信的多跳自组织网络。传感器网络能够扩展人们与遥远真实世界交互的能力。无线传感器网络是一种信息存取的新型平台。

在政府倡导的工业化和信息化结合的背景下,IoT 将是工业和其他信息业的真实启动点。一旦 IoT 大规模流行开来,更多的小物体需要安装智能传感器。用于动物、植物、机械和其他物品的传感器和电子标签将会大大超过现有电话的数量。根据目前对物联网的需求,近年需要大约数十亿的传感器和电子标签。专家估计到 2020 年,嵌入式芯片、传感器、无线射频和其他"智能物体"的数量将会超过 1 万亿。

虽然今天多数医院都有很多智能医疗设备,但这些设备并不能彼此通信。一旦这些设备能通过物联网完全连接起来,健康医疗的面貌将会发生深刻的变化。例如,IoT 会通过亚马逊的 Alexa 和智能电话在病人即将发生中风和心脏病之前提醒在家的病人有关其血凝块的情况。连接到电子医疗记录仪的传感器允许 IoT 快速诊断病人的身体状态以便帮助急诊医护人员并加快治疗。

另一有可能被彻底改变的重要行业是交通运输业。还在 5 年前,我们多数人还认为无人驾驶汽车的概念是科幻小说,或者至少是几十年后的事。然而,当谷歌在开放道路上测试其自动驾驶汽车时无人汽车已成为现实,传统汽车制造商如奔驰、尼桑和奥迪围绕这一运输游戏搅局者正在改变其商业策略。因为每一辆无人驾驶汽车都是一个智能传感器设备的集合,这些设备彼此连接成为一个整体网络,IoT 能够聚集并利用分布在整个网络中的集体智能同时驾驶汽车。

在制造业常会遇到工业互联网 IIoT 这一术语,它指的是 IoT 的工业部分应用。制造业的 IIoT 可能产生非常多的商业价值,这些最终会导致第四次工业革命,也就是所谓的工业 4.0。据估计未来一些成功的公司会利用 IoT 创造新的商业模式和提高生产率从而增加收入。到 2030 年,通过实施 IIoT 的潜在增长会产生 12 万亿美元的全球 GDP。

IoT 设备是家庭自动化的一大组成部分,包括照明设备、取暖和空调设备、媒体和安全系统。长远来看,通过自动关闭照明和电子设备可以带来节能的好处。

智能家庭或自动家庭可以以控制智能设备和电气的平台或集线器为基础。例如，使用苹果公司的 Homekit，制造商可以用 iOS 设备上的应用程序来控制家用产品，如苹果手机和苹果手表。这可以是专用 App 或 iOS 自带的应用如 Siri。也有专用的智能集线器是作为独立平台来连接不同的智能家居产品，这些包括 Amazon Echo、Google Home、苹果公司的 HomePod 和三星公司的 SmartThings Hub。

智能家居的另一重要应用是向那些残疾人士和老年人提供帮助。这些家居系统使用辅助技术来适应残疾人士的特定需要。声音控制可以帮助那些视力受限和移动受限的人士，而告警系统可以直接连到听力障碍人士佩戴的人工耳蜗。它们也可以配备其他的安全特性，包括监控紧急情况如跌倒和癫痫的传感器。这样应用的智能家居技术可以为用户提供更加自由和高质量的生活品质。

UNIT 10

PASSAGE

AI[1]

In the field of computer science, artificial intelligence (AI), sometimes called machine intelligence, is intelligence demonstrated by machines, in contrast to the natural intelligence displayed by humans and other animals.[2] Kaplan and Haenlein define AI as "a system's ability to correctly interpret external data, to learn from such data, and to use those learnings to achieve specific goals and tasks through flexible adaptation". Colloquially, the term "artificial intelligence" is applied when a machine mimics "cognitive" functions that humans associate with other human minds, such as "learning" and "problem solving".[3]

The current excitement about artificial intelligence (AI), particularly machine learning (ML), is palpable and contagious. The expectation that AI is poised to "revolutionize", perhaps even take over humanity, has elicited prophetic visions and concerns from some luminaries.[4] There is also a great deal of interest in the commercial potential of AI, which is attracting significant sums of venture capital and state-sponsored investment globally, particularly in China.[5] McKinsey, for instance, predicts the potential commercial impact of AI in several domains, envisioning markets worth trillions of dollars. All this is driven by the sudden, explosive, and surprising advances AI has made in the last 10 years or so. AlphaGo, autonomous cars, Alexa, Watson and other such systems, in game playing, robotics, computer vision, speech recognition, and natural language processing are indeed stunning advances.

But, as with earlier AI breakthroughs, such as expert systems in the 1980s and neural networks in the 1990s, there is also a considerable hype and a tendency to overestimate the promise of these advances, as market research firm Gartner and others have noted about emerging technology.

The implication is that AI could eventually end up doing all "things" that humans do, and do them much better—that is, achieve super-human performance as witnessed recently with AlphaGO and AlphaGO Zero.[6] Historically, the term AI reflected collectively to the following branches:

- Game playing — for example, Chess, Go;
- Symbolic reasoning and theorem proving — for example, Logic Theorist, MACSYMA;

- Robotics — for example, self-driving cars;
- Vision — for example, facial recognition;
- Speech recognition, natural language processing — for example, Siri;
- Distributed & evolutionary AI — for example, drone swarms;
- Hardware for AI — for example, Lisp machines;
- Expert systems or knowledge-based systems — for example, MYCIN, CONPHYDE.

Some of these are application-focused, such as game playing and vision. Others are methodological, such as expert systems and ML — the two branches that are most directly and immediately applicable to our domain.[7]

Many tasks in these different branches of AI share certain common features. They all require pattern recognition, reasoning, and decision-making under complex conditions. And they often deal with ill-defined problems, noisy data, model uncertainties, combinatorially large search spaces, nonlinearities and the need for speedy solutions.

Looking back some 30 years from now, history would recognize that there were three early milestones in AI.[8] One is Deep Blue defeating Gary Kasparov in chess in 1997, the second Watson becoming Jeopardy champion in 2011, and the third is the surprising win by AlphaGO in 2016. The AI advances that made these amazing feats possible are now poised to have an impact that goes far beyond game playing.[9]

AI research really started with a conference at Dartmouth College in 1956. It was a month long brainstorming session attended by many people with interests in AI. At the conference they wrote programs that were amazing at the time, beating people at checkers or solving word problems.[10] The Department of Defense started giving a lot of money to AI research and labs were created all over the world. Unfortunately, researchers really underestimated just how hard some problems were. The tools they had used still did not give computers things like emotions or common sense. Funding for AI research was cut, starting an "AI winter" where little research was done.[11]

AI research revived in the 1980s because of the popularity of expert systems, which simulated the knowledge of a human expert.[12] Expert systems, also called knowledge-based systems, rule-based systems, or production systems, are computer programs that mimic the problem-solving of humans with expertise in a given domain. By 1985, 1 billion dollars were spent on AI. New, faster computers convinced U. S. and British governments to start funding AI research again. However, the market for Lisp machines collapsed in 1987 and funding was pulled again, starting an even longer AI winter.[13]

AI revived again in the 90s and early 2000s with its use in data mining and medical diagnosis. This was possible because of faster computers and focusing on solving more specific problems. The excitement about expert systems waned in the 1990s due to these practical difficulties.

Although heuristic search and expert system proved suitable to solve well-defined, logical problems, such as playing chess, it turned out to be intractable to figure out explicit rules for solving more complex, fuzzy problems, such as image classification, speech recognition, or language translation.[14] Another boom of AI arose to take their place: machine learning (ML). ML is not a new concept. In 1959, Arthur Samuel, one of the pioneers of ML, defined machine learning as a "field of study that gives computers the ability to learn without being explicitly programmed".[15] That is, ML programs have not been clear entered into a computer, like the if-then statements above. ML programs, in a sense, adjust themselves in response to the data they're exposed to.

Deep learning is a kind of machine learning, a technology that enables computer systems to improve from experience and data.[16] Deep learning is a specific type of machine learning that has the power and flexibility to represent the vast universe as a system of nested hierarchical concepts (complex concepts defined by connections between simpler concepts, generalized to higher-level abstract representations).[17]

Artificial intelligence is a broad and active area of research, but it's no longer the sole province of academics; increasingly, companies are incorporating AI into their products. Google has been a pioneer in the use of machine learning — computer systems that can learn from data, as opposed to blindly following instructions. In particular, the company uses a set of machine-learning algorithms, collectively referred to as "deep learning", that allow a computer to do things such as recognize patterns from massive amounts of data.

NEW WORDS AND PHRASES

artificial　*a*.人工的,人造的
intelligence　*n*.智力,智能
demonstrate　*v*.证明,展示
interpret　*v*.解释,翻译,说明
external　*a*.外部的,表面的
adaptation　*n*.适应,改编
colloquially　*ad*.口语的,用通俗语
contagious　*a*.感染性的,会蔓延的
elicit　*vt*.抽出,引出
prophetic　*a*.预言的,预示的
luminary　*n*.发光体,杰出人物
domain　*n*.领域,域名
envision　*vt*.想象,预想
robotics　*n*.机器人学
recognition　*n*.认出,识别

stun	v.使震惊,打昏
neural	a.神经的,神经系统的
hype	n.大肆宣传
note	n.笔记,注解
methodological	a.方法的,方法论的
pattern	n.模式,图案
nonlinearity	n.非线性,非线性特征
speedy	a.快的,迅速的
session	n.会话,会议
amaze	v.使吃惊
checker	n.检验员,棋子
popularity	n.普及,流行
simulate	v.模仿,假装
mimic	v.模仿,模拟
diagnosis	n.诊断,分析
wane	v.衰落,变小
heuristic	a.启发的,探索的
intractable	a.棘手的,难治的
explicit	a.明确的,清楚的
fuzzy	a.模糊的,失真的
boom	v.使兴旺,急速发展
province	n.省,领域
academics	n.学术水平,学术知识
algorithm	n.算法,运算法则

NOTES

1. AI,人工智能。

2. sometimes called machine intelligence 为插入语;过去分词 demonstrated 和 displayed 做后置定语修饰前面的 intelligence;in contrast to,与……形成对照。本句译文:在计算机科学领域,与人类和其他动物显示的自然智能相比,人工智能 AI(有时被称为机器智能)是由机器表现出的智能。

3. colloquially 副词放在句首有强调作用;本句是被动句,可以按主动句翻译;that 引导定语从句,修饰前面的 functions;such as,比如。本句译文:通俗地说,"人工智能"一词是指机器模仿人类与其他人类思维相关的"认知"功能,比如"学习"和"解决问题"。

4. that AI is poised to "revolutionize"是由 that 引导的同位语从句,说明 expectation 的内容。本句译文:人们期待 AI 具有革命性的作用,甚至可以取代人类,这引起一些大人物的关注并预言了其愿景。

5. which 引导非限定性定语从句;venture capital,风险投资。本句译文:对 AI 商业潜力的兴趣吸引了全球特别是中国的大量风险投资和国家赞助的资金投入。

6. that 引导表语从句,做系动词 is 的表语;end up doing,以……而告终;that humans do 为定语从句,修饰前面的 things。本句译文:AI 所蕴含的最终能力可能会终结人类所做的所有事情,并且会比人类做得更好,也就是说能实现超人的性能,就像近来阿尔法狗和阿尔法狗零所见证的那样。

7. such as,例如;that 引导定语从句,修饰前面的 branches。本句译文:其他的在于方法论,例如专家系统和机器学习,这两个分支可最直接地用于我们的研究领域。

8. Looking back some 30 years from now 分词短语表示伴随情况;that 引导宾语从句,做 recognize 的宾语。本句译文:回望 AI 的 30 年历程,可以确认三件里程碑式的早期事件。

9. 本句是被动句;that made these amazing feats possible 为定语从句,修饰 advances;that goes far beyond game playing 定语从句,修饰前面的 impact。本句译文:人工智能的发展使这些惊人的壮举成为可能,现在它的影响已经远远超出了游戏的范畴。

10. that 引导定语从句,修饰 programs;at the time,当时,在那时;beating people at checkers 和 solving word problems 为分词短语,表示伴随情况。本句译文:在会议上,他们编写了当时令人惊叹的程序,击败了西洋跳棋玩家或解决了字符问题。

11. 本句是被动句;starting 引出分词短语,表示伴随情况;where 引导定语从句,修饰前面的 winter。本句译文:于是为 AI 研究所提供的资金消减了,所进行的 AI 研究很少,AI 的冬天开始了。

12. which 引导非限定性定语从句。本句译文:因为专家系统的流行,人工智能研究在 20 世纪 80 年代又开始兴起,专家系统模拟了人类专家的知识。

13. 由 and 连接两个并列句;starting an even longer AI winter 为分词短语,表示伴随情况。本句译文:然而,Lisp 机器的市场在 1987 年崩溃,资金再次被抽走,开始了 AI 更长的冬天。

14. although,虽然;such as,例如,比如;it 是形式主语,真正的主语是后边的不定式短语 to figure out explicit rules。本句译文:虽然启发式搜索和专家系统适合解决如下棋这样明确定义的逻辑问题,但是却很难处理更复杂和模糊的问题,如图像分类、语音识别或者语言翻译。

15. that 引导定语从句,修饰 study。本句译文:在 1959 年,ML 的先驱者之一 Arthur Samuel 将机器学习定义为:使计算机不必经过明确编程就能够学习的研究领域。

16. a kind of,一种;that 引导定语从句,修饰 technology。本句译文:深度学习是机器学习的一种,一种能够使计算机系统从经验和数据中得到提高的技术。

17. that 引导定语从句,对 machine learning 进行说明;to represent 不定式短语表示目的。本句译文:深度学习是一种特定类型的机器学习,具有强大的能力和灵活性,它将大千世界表示为嵌套的层次概念体系(由较简单概念间的联系定义复杂概念、从一般抽象概括到高级抽象)。

EXERCISES

一、请将下述词组译成英文

1. 人工智能　　2. 计算机科学领域　　3. 模仿认知功能
4. AI的商业潜力　　5. 比如专家系统　　6. 自然语言处理
7. 模拟人类解决问题　　8. 机器学习程序　　9. 活跃的研究领域
10. 向数据学习　　11. 深度学习　　12. 机器学习的特点类型
13. 高级抽象表现　　14. 一套机器学习算法

二、请将下述词组译成中文

1. the intelligence demonstrated by machines
2. in contrast to the natural intelligence
3. a system's ability to correctly interpret external data
4. to achieve specific goals and tasks
5. the potential commercial impact of AI in several domains
6. the speech recognition, and natural language processing
7. the tasks in these different branches of AI
8. to require pattern recognition, reasoning, and decision-making
9. an impact that goes far beyond game playing
10. because of the popularity of expert systems
11. the programs that were amazing at the time
12. to simulate the knowledge of a human expert
13. the use in data mining and medical diagnosis
14. to figure out explicit rules
15. such as image classification, speech recognition, or language translation
16. to solve well-defined, logical problems

三、选择合适的答案填空

1. Kaplan and Haenlein define AI as "a system's ability to correctly interpret external data, to learn from such data, and ________ those learnings to achieve specific goals and tasks through flexible adaptation".

A. to use　　B. using
C. used　　D. use

2. McKinsey, for instance, predicts the potential commercial impact of AI in several domains, ________ markets worth trillions of dollars.

A. envision　　B. to envision
C. envisioning　　D. envisioned

3. But, as with earlier AI breakthroughs, such as expert systems in the 1980s and neural networks in the 1990s, there is also considerable hype and a tendency to overestimate the promise of these advances, as market research firm Gartner and others

have noted about ________ technology.

A. emerge　　B. to emerge

C. emerged　　D. emerging

4. The implication is that AI could eventually end up doing all "things" that humans do, and do them much better — that is, achieve super-human performance as ________ recently with AlphaGO and AlphaGO Zero.

A. witnessing　　B. witnessed

C. to witness　　D. witness

5. And they often deal with ________ problems, noisy data, model uncertainties, combinatorially large search spaces, nonlinearities, and the need for speedy solutions.

A. ill-defining　　B. ill-defined

C. ill-define　　D. to ill-define

6. In 1959, Arthur Samuel, one of the pioneers of ML, defined machine learning as a "field of study that gives computers the ability to learn without ________ explicitly programmed".

A. to be　　B. been

C. be　　D. being

7. Expert systems, also called knowledge-based systems, rule-based systems, or production systems, are computer programs that mimic the problem-solving of humans with expertise in a ________ domain.

A. given　　B. to give

C. giving　　D. give

四、根据课文内容选择答案

1. In the field of computer science, artificial intelligence (AI), sometimes called machine intelligence, is ________, in contrast to the natural intelligence displayed by humans and other animals.

A. intelligence demonstrated by machines

B. the excitement about artificial intelligence

C. those learnings to achieve specific goals

D. significant sums of venture capital

2. The expectation that AI is poised to "revolutionize", perhaps even take over humanity, has ________ from some luminaries.

A. a machine mimics "cognitive" functions

B. the natural intelligence displayed by humans

C. the commercial impact of AI in several domains

D. elicited prophetic visions and concerns

3. McKinsey, for instance, predicts the potential commercial impact of AI in several domains, ________.

A. applied when a machine mimics "cognitive" functions

B. envisioning markets worth trillions of dollars

C. perhaps even take over humanity

D. attracting significant sums of venture capital

4. But, as with earlier AI breakthroughs, such as expert systems in the 1980s and neural networks in the 1990s, there is also considerable hype and a tendency to overestimate ________, as market research firm Gartner and others have noted about emerging technology.

A. the central dogma of AI

B. the potential commercial impact

C. the promise of these advances

D. the current excitement about artificial intelligence

5. The implication is that AI could eventually end up ________, and do them much better—that is, achieve super-human performance as witnessed recently with AlphaGO and AlphaGO Zero.

A. envisioning markets worth trillions of dollars

B. doing all "things" that humans do

C. looking back some 30 years from now

D. funding for AI research

6. Deep learning is ________ that has the power and flexibility to represent the vast universe as a system of nested hierarchical concepts (complex concepts defined by connections between simpler concepts, generalized to higher-level abstract representations).

A. a specific type of machine learning

B. perhaps even take over humanity

C. the need for speedy solutions

D. the popularity of expert systems

五、请将下列短文译成中文

1. Kaplan and Haenlein define AI as "a system's ability to correctly interpret external data, to learn from such data, and to use those learnings to achieve specific goals and tasks through flexible adaptation". Colloquially, the term "artificial intelligence" is applied when a machine mimics "cognitive" functions that humans associate with other human minds, such as "learning" and "problem solving".

2. McKinsey, for instance, predicts the potential commercial impact of AI in several domains, envisioning markets worth trillions of dollars. All this is driven by the sudden, explosive, and surprising advances AI has made in the last 10 years or so. AlphaGo, autonomous cars, Alexa, Watson, and other such systems, in game playing, robotics, computer vision, speech recognition, and natural language processing are

indeed stunning advances.

3. The implication is that AI could eventually end up doing all "things" that humans do, and do them much better — that is, achieve super-human performance as witnessed recently with AlphaGO and AlphaGO Zero. This implication is sometimes called the central dogma of AI. Historically, the term AI reflected collectively to the following branches.

4. Many tasks in these different branches of AI share certain common features. They all require pattern recognition, reasoning, and decision-making under complex conditions. And they often deal with ill-defined problems, noisy data, model uncertainties, combinatorially large search spaces, nonlinearities and the need for speedy solutions.

5. Looking back some 30 years from now, history would recognize that there were three early milestones in AI. One is Deep Blue defeating Gary Kasparov in chess in 1997, the second Watson becoming Jeopardy champion in 2011, and the third is the surprising win by AlphaGO in 2016. The AI advances that made these amazing feats possible are now poised to have an impact that goes far beyond game playing.

6. Although heuristic search and expert system proved suitable to solve well-defined, logical problems, such as playing chess, it turned out to be intractable to figure out explicit rules for solving more complex, fuzzy problems, such as image classification, speech recognition, or language translation. Another boom of AI arose to take their place: machine learning (ML).

7. ML is not a new concept. In 1959, Arthur Samuel, one of the pioneers of ML, defined machine learning as a "field of study that gives computers the ability to learn without being explicitly programmed". That is, ML programs have not been clear entered into a computer, like the if-then statements above. ML programs, in a sense, adjust themselves in response to the data they're exposed to.

8. Artificial intelligence is a broad and active area of research, but it's no longer the sole province of academics; increasingly, companies are incorporating AI into their products. Google has been a pioneer in the use of machine learning — computer systems that can learn from data, as opposed to blindly following instructions. In particular, the company uses a set of machine-learning algorithms, collectively referred to as "deep learning", that allow a computer to do things such as recognize patterns from massive amounts of data.

参考译文

AI

在计算机科学领域,与人类和其他动物显示的自然智能相比,人工智能AI(有时被称为机器智能)是由机器表现出的智能。Kaplan 和 Haenlein 将 AI 定义为“一个系统正确解释外部数据的能力,从这些数据中学习的能力,以及通过灵活的适应来利用这些学习以实现特定目标和任务的能力”。通俗地说,“人工智能”一词是指机器模仿人类与其他人类思维相关的“认知”功能,比如“学习”和“解决问题”。

很明显当前的人工智能特别是机器学习是令人激动且影响广泛。人们期待 AI 具有革命性的作用,甚至可以取代人类,这引起一些大人物的关注并预言了其愿景。对 AI 的商业潜力的兴趣吸引了全球特别是中国的大量风险投资和国家赞助的资金投入。例如麦肯锡公司预测 AI 在一些领域潜在的商业影响市场价值上万亿美元。所有这些都是由于 AI 在过去十几年突然取得的令人惊讶的进步所带来的。阿尔法狗、自动驾驶汽车、Alexa、Watson 以及其他类似的系统,在博弈、机器人、计算机视觉、语音识别和自然语言处理等方面都取得了令人惊异的进步。

但是,就像 AI 早期的突破一样,如 80 年代的专家系统和 90 年代的神经网络,对 AI 进行了过度的宣传且高估了这些进展的前景,正如市场调研公司 Gartner 和其他机构对新兴技术的记录那样。

AI 所蕴含的最终能力可能会终结人类所做的所有事情,并且会比人类做得更好,也就是说能实现超人的性能,就像近来阿尔法狗和阿尔法零所见证的那样。从历史发展来看,术语 AI 集中反映在下列方面:

- 游戏,例如国际象棋、围棋;
- 符号推理和理论证明,如逻辑理论家、MACSYMA(数学专家系统);
- 机器人,如自动驾驶汽车;
- 视觉,如人脸识别;
- 语音识别、自然语言处理,如 Siri;
- 分布式和渐进式 AI,如无人机群;
- 用于 AI 的硬件,如 Lisp 机器;
- 专家系统或基于知识的系统,如 MYCIN、CONPHYDE.

其中一些集中在应用上,如游戏和视觉分析。其他的在于方法论,例如专家系统和机器学习,这两个分支可最直接地用于我们的研究领域。

AI 不同分支的许多任务有共同的特性。他们都需要复杂条件下的模式识别、推理和决策。而且,他们经常要处理不明确的问题、噪声数据、模式不确定性、组合上的大搜索空间、非线性以及快速解决方案的需要。

回望 AI 的 30 年历程,可以确认三件里程碑式的早期事件。第一件是深蓝在 1997 年国际象棋比赛中击败世界冠军卡斯帕罗夫,第二件是 IBM 超级计算机 Watson 在 2011 年成为美国游戏节目 Jepardy 的冠军,第三件是 2016 年阿尔法狗在围棋比赛中令人惊讶的获胜。人工智能的发展使这些惊人的壮举成为可能,现在它的影响已经远远超出了游戏的范畴。

人工智能研究真正开始于 1956 年在达特茅斯学院举行的一次会议。这是一个为期一个月的头脑风暴会议,与会的许多人都是对 AI 感兴趣的。在会议上,他们编写了当时令人惊叹的程序,击败了西洋跳棋玩家或解决了字符问题。国防部开始为人工智能研究拨款,世界各地创建了很多AI实验室。不幸的是,研究人员真的低估了一些问题的难度。他们使用的工具还没能为计算机提供诸如情感或常识之类的能力。于是为AI研究所提供的资金消减了,所进行的AI研究很少,AI的冬天开始了。

因为专家系统的流行,人工智能研究在20世纪80年代又开始兴起,专家系统模拟了人类专家的知识。专家系统也称作基于知识的系统、基于规则的系统或生产系统,专家系统是计算机程序,它模仿在一特定领域具有专门知识的人怎样解决问题。到1985年,人工智能花费了10亿美元。新型的、更快的计算机使美国和英国政府再次开始资助人工智能研究。然而,Lisp机器的市场在1987年崩溃,资金再次被抽走,开始了AI更长的冬天。

人工智能在20世纪90年代和21世纪初期再次复活,用于数据挖掘和医学诊断。这可能是因为计算机速度更快,并专注于解决更具体的问题。由于一些实际的困难,在20世纪90年代人们对专家系统的热情冷淡下来。

虽然启发式搜索和专家系统适合解决如下棋这样明确定义的逻辑问题,但是却很难处理更复杂和模糊的问题,如图像分类、语音识别或者语言翻译。于是出现了AI的另一个繁荣:机器学习ML。ML不是一个新概念。在1959年,ML的先驱者之一Arthur Samuel将机器学习定义为:使计算机不必经过明确编程就能够学习的研究领域。也就是,ML程序(像if-then语句)没有明确地输入计算机。某种意义上,ML程序面对数据时能够相应地调整自己。

深度学习是机器学习的一种,一种能够使计算机系统从经验和数据中得到提高的技术。深度学习是一种特定类型的机器学习,具有强大的能力和灵活性,它将大千世界表示为嵌套的层次概念体系(由较简单概念间的联系定义的复杂概念、从一般抽象概括到高级抽象表示)。

人工智能是一个广泛而活跃的研究领域,但它不再是学术界的唯一领域,越来越多的公司将人工智能应用到他们的产品中。谷歌一直是使用机器学习的先锋,与盲目地按照指令学习不同,机器学习是一种能够向数据学习的计算机系统。特别是公司使用了一些称作深度学习的机器学习算法,这些算法允许计算机从大量的数据中进行模式识别。

UNIT 11

PASSAGE

IP Version 6[1]

As the Internet began to grow at a dramatic rate during the late 1980s and early 1990s, engineers realized that the current version of the IP protocol would not be adequate to meet the demands of the Internet's growth.[2] Of particular concern was the availability of adequate IP addresses for all kinds of devices accessing the Internet. With 32 bits, IPv4 can theoretically provide up to 4.2 billion addresses. However, even with uniform allocation, this is insufficient. Allocation policies in practice make this worse, a few organizations have an overabundance of addresses, whereas many others have ended up with too few. According to the report released by Gartner, there will be 25 billion IoT devices by 2020, and a large number of perceptive terminals or machines will be connected to the mobile network or the Internet, which will inevitably cause a huge demand for IP addresses.[3] IPv6 is the fundamental way to solve the address problem. The designers of IPv6 have taken full account of the existing advantages of IPv4 and made a lot of improvements and extensions to make it more powerful and efficient than IPv4's processing performance.

IPv6 also provides additional features. Address auto-configuration allows hosts to autoconfigure their IP addresses without the need for a centralized server. The header, which contains the essential fields for the IP protocol to work, has been simplified for more efficient processing; even with 40 bytes, the IPv6 header is more amenable to header compression than the IPv4 header.[4] Extensions to the base IPv6 header provide better support for security and mobility, the mobility header in particular has been designed to support the mobile IP protocol and its enhancements.

With the basic addressing support and additional support, especially for security and mobility, the new IPv6 protocol can form the basis of the next-generation Internet. It is, hence, important to understand some of its basic operations.

The Internet protocols often define standards for hosts and routers to communicate. IPv6 is one such important protocol. In order for a standard to be implementable as well as interoperable, it needs to define unambiguous messages that the protocol uses.[5] The messages carry information for the IP software to interpret, as well as "payload" a user generated such as web pages, streamed content and VoIP. The information carried for the IP software can be generally referred to as the "header", which specifies how the message must be interpreted and processed.[6] The header definition and semantics must be unambiguous so that the sender and the receiver can communicate. A header "format" does just that. It allows a sender to construct and fill the fields in the header in a way that the receiver can understand and interpret. IPv6 packet consists of three parts: header, extended header and upper protocol data unit. The fixed header contains eight fields with a fixed total length of 40 bytes. The following section provides a description of the IPv6 header.

The header in an IPv6 packet is shown in Fig. 11-1. The description of each of the header fields follows.

Each" + - " corresponds to 1 bit. Hence, the Version field consists of 4 bits. The Version field is set to 6 to indicate the protocol version.

The traffic class field actually consists of two subfields. The first 6bits of this field constitute the differentiated services code point (DSCP), and are used for providing different forwarding treatment (Quality of Service, QoS) to traffic. A source code may mark these bits to request certain "differentiated services", and a router may provide the corresponding forwarding behavior. These bits may be remarked by intermediate routers to ensure that the traffic is compliant to what is agreed upon.[7] The remaining two bits of the field are reserved for "Explicit Congestion Notification (ECN)", which is used to inform in advance the transport protocol of congestion along the path a packet takes.

The flow label is used to identify a flow, which is a sequence of packets from a source to a destination.[8] Traditionally, the flow classification has been based on the 5-tuple of source and destination IP addresses, source and destination ports, and the transport protocol type. Flow Label serves as an IP header field that can assist in flow classification at the intermediate nodes.[9]

The length of the rest of the packet following the IPv6 header is denoted by payload length. The type of the header immediately following the IPv6 header is identified by next header. For data, this is typically TCP or unreliable datagram protocol (UDP). However, IPv6 defines multiple extension header which may be present.

Hop limit determines how far a packet should traverse on the Internet. Its value is decremented by one by each node that forwards the packet. When the value reaches zero, the packet is decremented.

The source and destination addresses specify the originator and the receiver of the packet, respectively. The Destination Address does not specify the ultimate receiver if the Routing Header is present.

There are three types of addresses in IPv6: unicast, anycast, and multicast. Each of these addresses can have different scopes, which limits their applicability and usage. There are no broadcast addresses in IPv6.[10]

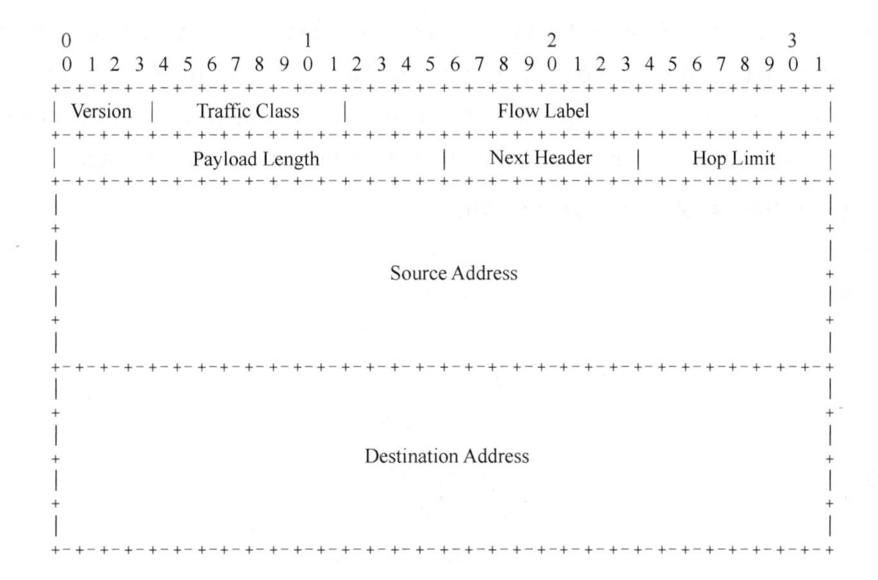

Fig. 11-1 IPv6 Header

The addresses are represented in text form using the notation "X:X:X:X:X:X:X:X", where each X is the hexadecimal value corresponding to 16 bits of the overall address.[11] An example is ABCD:1234:FEDC:0123:0:0:0:1. A contiguous string of zeros can be compressed using the notation "::". For example, the above address can be represented as ABCD:1234:FEDC:0123::1.

With the rapid development of Internet, IPv4 protocol can no longer meet the needs of users. This is mainly due to the limitations of IPv4 in terms of addresses, routing and security. Correspondingly, IPv6 has the advantage of large address space, security, mobility, quality of service and so on. So IPv6 protocol has become the inevitable trend of network development. However IPv4 and IPv6 are incompatible protocols, so a solution to transition is required. Because it is impossible to change their network to IPv6 at once, IPv4 and IPv6 will coexist for a few years. In order to achieve smooth and stepwise transition, IETF recommends three kinds of transition mechanisms: dual stack, tunneling and translation technology.

Dual stack mechanism includes two protocol stacks that operate in parallel and allow network nodes to communicate either via IPv4 or IPv6.[12] It is the foundation of other IPv6 transition mechanism.

The core of tunneling is that an IPv6 (IPv4) protocol packet is encapsulated in IPv4 (IPv6) protocol packet, then IPv6 (IPv4) packet can be transmitted across IPv4 (IPv6) network.[13] For example, IPv6 over IPv4 tunnel (IPv6 packet encapsulated in IPv4 packet) provides a method of using the existing IPv4 routing infrastructure to transfer IPv6 packet. IPv4 over IPv6 tunnel is just the opposite of IPv6 over IPv4.

Translation technology can be used to implement address and protocol translation between IPv4 and IPv6. So translation technology is a manner of mutual access between IPv4-only node and IPv6-only node. By using specific translation technology, original IPv4 user terminal does not require upgrading, because network equipments will accomplish all the work, including the translation of protocol and address between IPv4 and IPv6.[14]

NEW WORDS AND PHRASES

configuration	*n*. 配置,结构
host	*n*. 主机,主人
header	*n*. 页眉,数据头
field	*n*. 领域,字段
dramatic	*a*. 喜剧的,引人注目的
protocol	*n*. 协议,草案
availability	*n*. 可用性,有效性
overabundance	*n*. 过多,过于丰富
explosive	*a*. 爆炸的,爆炸性的
emergence	*n*. 出现,浮现
expose	*vt*. 揭露,揭发,使曝光
implementable	*a*. 可执行的
unambiguous	*a*. 不含糊的,清楚的
specify	*vt*. 指定,详细说明
semantics	*n*. 语义学,语义论
description	*n*. 描述,描写
forward	*v*. 发送,转寄
compliant	*a*. 顺从的,服从的
identify	*v*. 确定,识别
denote	*vt*. 表示,指示
hop	*n*. 单足短距离跳跃
traverse	*v*. 穿过,来回移动
decrement	*n*. 渐减,减缩
ultimate	*a*. 最终的,终极的
notation	*n*. 符号,乐谱,注释
hexadecimal	*a*. 十六进制的
contiguous	*a*. 连续的,邻近的
inevitable	*a*. 必然的,不可避免的
incompatible	*a*. 不相容的,矛盾的

coexist	*vi*. 共存，和平共处
stepwise	*a*. 逐步的，逐渐的
dual	*a*. 双的，双重的
stack	*n*. 一堆，堆栈
tunneling	*n*. 隧道效应，隧道技术
encapsulate	*v*. 压缩，将……封进内部

NOTES

1. 本文简要介绍了 IPv6 的相关技术。

2. that 引导宾语从句，做动词 realized 的宾语。本句译文：20 世纪 80 年代末期和 90 年代初期，互联网开始以惊人的速度发展，工程师们意识到当前 IP 协议的版本不足以满足互联网发展的需求。

3. according to，根据；released by Gartner 过去分词短语做后置定语，修饰前面的 report；which 引导限定性定语从句。本句译文：根据 Gartner 发布的报告，到 2020 年将有 250 亿个物联网设备出现，大量的感知终端或机器接入移动网或互联网，必将引发对 IP 地址的巨大需求。

4. which 引导非限定定语从句。本句译文：为了更高效地处理，包含 IP 协议工作的基本字段的报头已被简化；即使是 40 字节，IPv6 报头也比 IPv4 报头更适合报头压缩。

5. that the protocol uses 由 that 引导定语从句，修饰 messages。本句译文：为了使标准既可实现又可互操作，它需要定义协议使用的明确的消息。

6. carried for the IP software 过去分词短语做后置定语，修饰 information；which 引导定语从句；how the message must be interpreted and processed 由 how 引导宾语从句，做动词 specifies 的宾语；referred to as，被称作。本句译文：为 IP 软件携带的信息通常被称为“报头”，它指定了必须如何解释和处理消息。

7. that 引导宾语从句，做动词 ensure 的宾语。本句译文：这些位可能被中间路由器再次标记，以确保流符合所做的约定。

8. 本句是被动句；which 引导非限定定语从句。本句译文：流标签用于标识流，是从源到目的地的数据包序列。

9. that can assist in flow classification at the intermediate nodes 由 that 引导定语从句，修饰 field。本句译文：流标签作为一个 IP 报头字段，可以在中间节点帮助流分类。

10. which 引导非限定性定语从句。本句译文：每个地址都有不同的范围，限制了它们的适用性和使用。

11. using the notation“$X:X:X:X:X:X:X:X$”为现在分词短语，表示伴随情况；where 引导定语从句；corresponding to，对应于。本句译文：这些地址用“$X:X:X:X:X:X$”的符号形式来表示，其中每个 X 对应整个 16 位地址的十六进制值。

12. that 引导定语从句，修饰 stacks。本句译文：双栈机制包括两个并行操作的协议栈，允许网络节点通过 IPv4 或 IPv6 进行通信。

13. that引导表语从句,做系动词is的表语。本句译文:隧道化的核心是将一个IPv6(IPv4)协议包封装在IPv4(IPv6)协议包中,然后IPv6(IPv4)包可以通过IPv4(IPv6)网络进行传输。

14. because引导原因状语从句;including the translation of protocol and address between IPv4 and IPv6为现在分词短语,表示伴随情况。本句译文:通过使用特定的转换技术,原有的IPv4用户终端无须升级,包括IPv4和IPv6之间协议和地址的转换在内的所有工作都将由网络设备完成。

EXERCISES

一、请将下述词组译成英文

1. 以惊人的速度增长
2. 当前IP协议版本
3. 包括两个协议栈
4. 基本的IPv6报头
5. 双栈机制
6. 标识一个流
7. 足够的IP地址
8. 报头类型
9. 填充报头中的字段
10. 大量的地址
11. 基本的地址支持
12. 解决地址问题
13. IPv4现有的优点
14. IPv4的处理性能
15. 为主机定义标准
16. 相连的零串

二、请将下述词组译成中文

1. the address auto-configuration
2. the need for a centralized server
3. to support the mobile IP protocol
4. to define unambiguous messages
5. the source and destination IP addresses
6. to indicate the protocol version
7. the essential fields for the IP protocol
8. to meet the demands of the Internet's growth
9. to provide better support for security and mobility
10. the basis of the next-generation Internet
11. to provides a description of the IPv6 header
12. the advantage of large address space
13. the inevitable trend of network development
14. to change their network to IPv6 at once
15. in order to achieve smooth and stepwise transition
16. to implement address and protocol translation
17. the hexadecimal value corresponding to 16 bits of the overall address

三、选择合适的答案填空

1. Of particular concern was the availability of adequate IP addresses for all kinds of devices ________ the Internet.

A. accessing　　B. access

C. to access　　D. accessed

2. The designers of IPv6 have taken full account of the existing advantages of IPv4 and made a lot of improvements and extensions to make it more powerful and efficient than IPv4's ________ performance

A. process　　B. to process

C. processed　　D. processing

3. With the basic addressing support and additional support, especially for security and mobility, the new IPv6 protocol can form the basis of the next-generation Internet. It is, hence, important ________ some of its basic operations.

A. understand　　B. to understand

C. understanding　　D. understood

4. The messages carry information for the IP software to interpret, as well as "payload" a user ________ such as web pages, streamed content and VoIP.

A. to generate　　B. generating

C. generated　　D. generate

5. A header "format" does just that. It allows a sender ________ and fill the fields in the header in a way that the receiver can understand and interpret.

A. construct　　B. constructing

C. to construct　　D. constructed

6. The length of the rest of the packet ________ the IPv6 header is denoted by payload length.

A. to follow　　B. following

C. followed　　D. follow

四、根据课文内容选择答案

1. However, even with uniform allocation, this is insufficient. Allocation policies in practice make this worse; a few organizations have an overabundance of addresses, whereas ________.

A. the IP protocol would not be adequate

B. many others have ended up with too few

C. address auto-configuration allows hosts to autoconfigure their IP addresses

D. the Internet protocols often define standards for hosts

2. With the basic addressing support and additional support, especially for ________, the new IPv6 protocol can form the basis of the next-generation Internet.

A. the existing advantages of IPv4

B. a lot of improvements

C. the header fields

D. security and mobility

3. In order for a standard to be implementable as well as interoperable, it ________

that the protocol uses.

A. needs to define unambiguous messages

B. provide better support for security and mobility

C. can understand and interpret

D. consists of two subfields

4. A header "format" does just that. It allows a sender to construct and fill the fields in the header in a way that ________ and interpret.

A. IPv6 also provides additional features

B. the Version field consists of 4 bits

C. the receiver can understand

D. allocation policies in practice make this worse

5. The flow label ________, which is a sequence of packets from a source to a destination.

A. is used to identify a flow

B. is set to 6 to indicate the protocol version

C. is used to inform in advance the transport protocol

D. is denoted by payload length

6. The source and destination addresses ________, respectively.

A. understand some of its basic operations

B. fill the fields in the header in a way

C. serves as an IP header field

D. specify the originator and the receiver of the packet

7. With the rapid development of Internet, IPv4 protocol ________.

A. achieve smooth and stepwise transition

B. can no longer meet the needs of users

C. includes two protocol stacks

D. does not require upgrading

五、请将下列短文译成中文

1. As the Internet began to grow at a dramatic rate during the late 1980s and early 1990s, engineers realized that the current version of the IP protocol would not be adequate to meet the demands of the Internet's growth. Of particular concern was the availability of adequate IP addresses for all kinds of devices accessing the Internet.

2. According to the report released by Gartner, there will be 25 billion IoT devices by 2020, and a large number of perceptive terminals or machines will be connected to the mobile network or the Internet, which will inevitably cause a huge demand for IP addresses. IPv6 is the fundamental way to solve the address problem.

3. IPv6 also provides additional features. Address auto-configuration allows hosts to autoconfigure their IP addresses without the need for a centralized server. The

header, which contains the essential fields for the IP protocol to work, has been simplified for more efficient processing; even with 40 bytes, the IPv6 header is more amenable to header compression than the IPv4 header.

4. With the rapid development of Internet, IPv4 protocol can no longer meet the needs of users. This is mainly due to the limitations of IPv4 in terms of addresses, routing and security. Correspondingly, IPv6 has the advantage of large address space, security, mobility, quality of service and so on. So IPv6 protocol has become the inevitable trend of network development.

5. With the basic addressing support and additional support, especially for security and mobility, the new IPv6 protocol can form the basis of the next-generation Internet. It is, hence, important to understand some of its basic operations. The Internet protocols often define standards for hosts and routers to communicate. IPv6 is one such important protocol.

6. The information carried for the IP software can be generally referred to as the "header", which specifies how the message must be interpreted and processed. The header definition and semantics must be unambiguous so that the sender and the receiver can communicate. A header "format" does just that.

7. A source code may mark these bits to request certain "differentiated services", and a router may provide the corresponding forwarding behavior. These bits may be remarked by intermediate routers to ensure that the traffic is compliant to what is agreed upon. The remaining two bits of the field are reserved for "Explicit Congestion Notification (ECN)", which is used to inform in advance the transport protocol of congestion along the path a packet takes.

8. The length of the rest of the packet following the IPv6 header is denoted by payload length. The type of the header immediately following the IPv6 header is identified by next header. For data, this is typically TCP or unreliable datagram protocol (UDP). However, IPv6 defines multiple extension header which may be present.

参考译文

IPv6

20世纪80年代末期和90年代初期，互联网开始以惊人的速度发展，工程师们意识到当前IP协议的版本不足以满足互联网发展的需求。特别令人关切的是为访问互联网的各种设备提供足够的IP地址可用性。基于32位地址的IPv4理论上可以提供多达42亿个地址。然而，即使有统一的地址分配，这也是不够的。实际上，分配政策让情况变得更糟，一些机构的地址过多，而其他许多机构的地址太少。根据Gartner发布的报告，到2020年将有250亿个物联网设备出现，大量的感知终端或机器接入移动网或互联网，必将引发对IP地址的巨大需求,而IPv6是解决地址问题的根本途径。IPv6的设计者充分考虑了IPv4现有的优点并进行了大量的改进和功能扩充,使其比IPv4处理性能更加强大、高效。

IPv6还提供了额外的一些功能。地址自动配置允许主机自动配置它们的IP地址,而不需要一个中央服务器。为了更高效地处理,包含IP协议工作的基本字段的报头已被简化;即使是40字节,IPv6报头也比IPv4报头更适合报头压缩。对基础IPv6报头的扩展向人们提供了对安全性和移动性的更好支持,移动性报头被特别设计来支持移动IP协议及其增强功能。

有了基本的寻址支持和额外的支持,特别是安全性和移动性,新的IPv6协议可以形成下一代互联网的基础。因此,理解它的一些基本操作很重要。

互联网协议通常为主机和路由器的通信定义标准。IPv6就是这样一个重要的协议。为了使标准既可实现又可互操作,它需要定义协议使用的明确的消息。这些消息携带着IP软件解释的信息,以及用户生成的"有效负载",如web页面、流内容和VoIP。为IP软件携带的信息通常被称为"报头",它指定了必须如何解释和处理消息。报头的定义和语义必须明确,以便发送方和接收方能够通信。报头"格式"就是这样做的。它允许发送方以接收方能够理解和解释的方式构造和填充报头中的字段。IPv6数据包由三部分构成,即报头、扩展报头以及上层协议数据单元。固定报头包含8个字段,总长度固定为40字节。下面的部分提供了对IPv6报头的描述。IPv6数据包的报头如图11-1(第135页)所示。下面是每个报头字段的描述。

每个"+—"对应于1位。因此,版本字段由4位组成。版本字段设置为6表示协议版本。

流量类别字段实际上由两个子字段组成。该字段的前6位构成差异化服务代码点(DSCP),用于对流量提供不同的转发处理(服务质量,QoS)。源代码可以标记这些位来请求某些"差异化服务",路由器可以提供相应的转发行为。这些位可能被中间路由器再次标记,以确保流符合所做的约定。字段的其余两位保留为"显示拥塞通知(ECN)",用于预先通知传输协议数据包所走路径上的拥塞情况。

流标签用于标识流,流是从源到目的地的数据包序列。传统上,流分类是基于源和目的IP地址、源和目的端口以及传输协议类型的五元组。流标签作为一个IP报头字段,可以在中间节点帮助流分类。

IPv6报头后面的包的其余长度由有效负载长度表示。IPv6报头之后的报头类型由下一个报头标识。对于数据而言,这通常是TCP或不可靠数据报协议UDP。然而,IPv6定义了可能存在的多个扩展头。

跳数限制决定了一个包在互联网上应该走多远。它的值由每个转发数据包的节点递减一个。当这个值达到0时,数据包就会递减。

源地址和目的地址分别指定包的发端者和接收者。如果存在路由报头,则目标地址不指定最终接收方。

IPv6中有三种类型的地址:单播、任意播和多播。每个地址都有不同的范围,限制了它们的适用性和使用。IPv6中没有广播地址。

这些地址用“X：X：X：X：X：X”的符号形式来表示，其中每个 X 对应整个16位地址的十六进制值。一个例子是ABCD：1234：FEDC：0123：0：0：1。用符号“：：”来压缩连续的零串。例如，上面的地址可以被表示为ABCD：1234：FEDC：0123：：I。

随着互联网的飞速发展，IPv4协议已经不能满足用户的需求。这主要是由于IPv4在地址、路由和安全性方面的限制。相应地，IPv6在大地址空间、安全性、移动性、服务质量等方面具有优势。因此IPv6协议已成为网络发展的必然趋势。然而，IPv4和IPv6是不兼容的协议，因此需要一个过渡的解决方案。因为不可能一下子把他们的网络改成IPv6，IPv4和IPv6将会共存几年。为了实现平稳、逐步的转换，IETF推荐了三种转换机制：双栈、隧道和转换技术。

双栈机制包括两个并行操作的协议栈，允许网络节点通过IPv4或IPv6进行通信。它是其他IPv6过渡机制的基础。

隧道化的核心是将一个IPv6（IPv4）协议包封装在IPv4（IPv6）协议包中，然后IPv6（IPv4）包可以通过IPv4（IPv6）网络进行传输。例如，IPv6在IPv4隧道（IPv6包封装在IPv4包中）提供了一种使用现有IPv4路由基础设施传输IPv6包的方法。IPv6隧道上的IPv4与IPv4上的IPv6正好相反。

转换技术可用于实现IPv4和IPv6之间的地址和协议转换。因此转换技术是一种IPv4节点和IPv6节点之间相互访问的方式。通过使用特定的转换技术，原有的IPv4用户终端无须升级，包括IPv4和IPv6之间协议和地址的转换在内的所有工作都将由网络设备完成。

UNIT 12

PASSAGE

Circuit Switching and Packet Switching[1]

There are two basic types of switching techniques: circuit switching and message switching. In circuit switching, a total path of connected lines is set up from the origin to the destination at the time the call is made, and the path remains allocated to the source-destination pair (whether used or not) until it is released by the communicating parties.[2] The switches, called circuit switches (or office exchange in telephone jargon), have no capability of storing or manipulating user's data on their way to the destination.[3] The circuit is set up by a special signaling message that finds its way through the network, seizing channels in the path as it proceeds.[4] Once the path is established, a return signal informs the source to begin transmission. Direct transmission of data from source to destination can then take place without any intervention on the part of the subnet.

In message switching, the transmission unit is a well-defined block of data called a message. In addition to the text to be transmitted[5], a message comprises a header and a checksum. The header contains information regarding the source and destination addresses as well as other control information[6]; the checksum is used for error control purposes. The switching element is a computer referred to as a message processor[7], with processing and storage capabilities. Messages travel independently and asynchronously, finding their own way from source to destination.[8] First the message is transmitted from the host to the message processor to which it is attached.[9] Once the message is entirely received, the message processor examines its header, and accordingly decides on the next outgoing channel on which to transmit it.[10] If this selected channel is busy, the message waits in a queue until the channel becomes free, at which time transmission begins. At the next message processor, the message is again received, stored, examined, and transmitted on some outgoing channel, and the same process continues until the message is delivered to its destination. This transmission technique is also referred to as the store-and-forward transmission technique.[11]

A variation of message switching is packet switching. Here the message is broken up into several pieces of a given maximum length, called packets. As with message switching[12], each packet contains a header and a checksum. Packets are transmitted independently in a store-and-forward manner.

With circuit switching, there is always an initial connection cost incurred in setting up the circuit.[13] It is cost-effective only in those situations where once the circuit is set up there is a guaranteed steady flow of information transfer to amortize the initial cost.[14] This is certainly the case with voice communication in the traditional way, and indeed circuit switching is the technique used in the telephone system. Communication among computers, however, is characterized as bursty. Burstiness is a result of the high degree of randomness encountered in the message-generation process and the message size, and of the low delay constraint required by the user.[15] The users and devices require the communication resources relatively infrequently; but when they do, they require a relatively rapid response. If a fixed dedicated end-to-end circuit were to be set up connecting the end users[16], then one must assign enough transmission bandwidth to the circuit in order to meet the delay constraint with the consequence that the resulting channel utilization is low.[17] If the circuit of high bandwidth were set up and released at each message transmission request, then the set-up time would be large compared to the transmission time of the message, resulting again in low channel utilization.[18] Therefore, for bursty users (which can also be characterized by high peak-to-average data rate requirements), store-and-forward transmission techniques offer a more cost-effective solution, since a message occupies a particular communications link only for the duration of its transmission on that link, the rest of the time it is stored at some intermediate message switch and the link is available for other transmissions. Thus the main advantage of store-and-forward transmission over circuit switching is that the communication bandwidth is dynamically allocated, and the allocation is done on the fine basis of a particular link in the network and a particular message (for a particular source-destination pair).

Packet switching achieves the benefits discussed so far and offers added features. It provides the full advantage of the dynamic allocation of the bandwidth, even when messages are long. Indeed, with packet switching, many packets of the same message may be in transmission simultaneously over consecutive links of a path from source to destination, thus achieving a "pipelining" effect and reducing considerably the overall transmission delay of the message as compared to message switching. It tends to require smaller storage allocation at the intermediate switches. It also has better error characteristics and leads to more efficient error recovery procedures, as it deals with smaller entities. Needless to say[19], packet switching presents design problems of its own, such as the need to reorder packets of a given message that may arrive at the destination node out of sequence.[20]

NEW WORDS AND PHRASES

destination — *n*. 目的地,终点

allocate	vt. 分配，分派，配给
release	vt. 释放，放松，发布
jargon	n. 行话，土语
manipulate	vt. 处理，操纵
seize	vt. 抓住，占领，俘获
proceed	vi. 进行，继续
inform	vt. 通知，告诉
intervention	n. 干预，干涉，妨碍
subnet	n. 子网，分网
block	n. 一批，一组，一块
header	n. 头，头部
checksum	n. 检验和，检验项
regard	v. 与……有关，涉及
attach	vt. 缚，系，捆
entirely	ad. 完全地，彻底地
accordingly	ad. 相应地
queue	n. 行列，长队
variation	n. 变动，变化，变更
incur	vt. 招致，惹起
guarantee	vt. 保证，担保
steady	a. 平稳的，不变的
amortize	vt. 缓冲，分摊
bursty	n. 突发性，突发
randomness	n. 随机，机遇
constraint	n. 强迫，强制，制约
relatively	ad. 相对地，比较地
response	n. 回答，响应
dedicated	a. 专用的
assign	vt. 分配，指派，委派
utilization	n. 利用
dynamically	ad. 动态地，有生气地
benefit	n. 利益，好处，恩惠
simultaneously	ad. 同时地，同时发生
consecutive	a. 连续的，连贯的
pipeline	v. 用管道输送

procedure　　n. 过程,步骤
entity　　n. 存在,实体

NOTES

1. 本篇课文涉及交换领域,题目可译成:电路交换与分组交换。

2. 本句中的 the call is made 是一个定语从句,修饰 the time,可译成:呼叫(被做出)的时候。allocated 是过去分词,在句中做形容词。

3. 句中的 storing 和 manipulating 均为动名词形式,user's data 是它们的动词宾语。

4. 句中的 seizing 为现在分词,它作为动词 finds 的伴随情况。

5. to be transmitted 为动词不定式的被动态,做定语,修饰 text,可译成:要传的内容。

6. regarding 是现在分词,修饰 information,可译为:有关信源和目的地地址的信息。

7. referred to as… 过去分词短语,做 computer 的定语,可译为:被称为报文处理器的计算机。

8. finding 为现在分词,作为动词 travel 的伴随情况。

9. to which 中的 which 代表 message processor,它引导一个定语从句,其中的 to 是动词 attach 的要求。to which it is attached 可译为:它被连接到的(报文处理器)。

10. on which 中的 which 代表前面的 channel。

11. store-and-forward,存储转发。

12. as with…,就像……一样。

13. incurred 是过去分词,做形容词,修饰 cost,可译成:(所)经历的……。

14. to amortize the initial cost 是动词不定式短语,表示目的,可译成:以便缓冲初始费用。

15. 在本句中,encountered in… size 是过去分词短语,做定语,修饰 the high degree of randomness,可译成:在……(方面)所遇到的高度随机性。

16. 句中 connecting the end users 为现在分词短语,该现在分词在句中做状语,表示目的。逗号前的整个句子可译为:如果要建立一个固定专用的端到端电路以连接两端的用户。

17. with the consequence that… 可译为:其结果为……。

18. resulting… 为现在分词短语,该短语在句中做状语,表示结果,可译成:亦造成了很低的电路利用率。

19. needless to say,不用说,当然。

20. need to reorder 中的动词不定式 to reorder 做 need 的定语,可译成:重新排序的必要。

EXERCISES

一、请将下述词组译成英文

1. 电路交换 2. 分组交换 3. 报文交换 4. 子网
5. 报头 6. 目的地址 7. 误差控制 8. 存储转发方式
9. 突发性 10. 传输时延 11. 中间交换设备 12. 交换技术
13. 返回信号 14. 报文处理机 15. 给定最大长度 16. 信息转移
17. 随机性 18. 专用电路 19. 电路利用率

二、请将下述词组译成中文

1. the capability of storing or manipulating user's data
2. the special signaling message
3. a well defined block of data called a message
4. the information regarding the source and destination addresses
5. the computer referred to as a message processor
6. the store-and-forward transmission technique
7. the dynamic allocation of the bandwidth
8. the overall transmission delay of the message
9. switching technique
10. circuit switching
11. message switching
12. packet switching
13. total path of connected lines
14. source-destination pair
15. communication parties
16. transmission unit
17. initial connection cost incurred in setting up the circuit
18. low delay constraint required by the user
19. the fixed dedicated end-to-end circuit
20. low channel utilization

三、选择合适的答案填空

1. The switch, ________ circuit switches, have no capability of ________ user's data on their way to the destination.

A. calling, stored B. called, storing
C. to call, to store D. be called, be stored

2. The circuit is set up by a special signaling message ________ finds its way through the network, ________ channels in the path as it proceeds.

A. that, seizing B. where, seizing
C. who, seized D. one, be seized

3. The switching element is a computer ________ a message processor, with processing and storage capabilities. Message travels independently and asynchronously, ________ their own way from source to destination.

A. to call, find B. calling, found

C. referred to as, finding D. be called, sending

4. A variation of message switching is packet switching. Here the message is broken up into several pieces of a ________ maximum length, ________ packets.

A. given, called B. give, call

C. giving, calling D. to give, to call

四、根据课文内容选择正确答案

1. ________, a total path of connected lines is set up from the origin to the destination at the time the call is made, and the path remains allocated to the source-destination pair until it is released by the communication parties.

A. In message switching

B. In packet switching

C. In circuit switching

D. In digital switching

2. In message switching, the transmission unit is a well defined block of data called a message. In addition to the text to be transmitted, a message comprises ________.

A. a start and end of data

B. a header and checksum

C. a start and a stop bit

D. a transmitter and receiver

3. A variation of message switching is packet switching. Here the message is broken up into several pieces of a given maximum length, ________.

A. called information

B. called texts

C. called packets

D. called messages

4. The header of a message contains information regarding ________; the checksum is used for error control purposes.

A. the text to be transmitted

B. the source and destination addresses

C. the voice channels

D. many packets between two computers

5. With circuit switching, there is always an initial connection cost incurred in setting up the circuit. It is cost-effective only in those situations where once the circuit is set up ________.

A. the voice conversation begins

B. there is no any intervention on the part of the subnet

C. there is a message to be transmitted

D. there is a guaranteed steady flow of information transfer

五、请将下述短文译成中文

1. In circuit switching, a total path of connected lines is set up from the origin to the destination at the time the call is made, and the path remains allocated to the source-destination pair (whether used or not) until it is released by the communicating parties. The switches, called circuit switches, have no capability of storing or manipulating user's data on their way to the destination. The circuit is set up by a special signalling message that finds its way through the network, seizing channels in the path as it proceeds. Once the path is established, a return signal informs the source to begin transmission.

2. In message switching, the transmission unit is a well defined block of data called a message. In addition to the text to be transmitted, a message comprises a header and a checksum. The header contains information regarding the source and destination addresses as well as other control information; the checksum is used for error control purposes. The switching element is a computer referred to as a message processor, with processing and storage capabilities.

3. Messages travel independently and asynchronously, finding their own way from source to destination. First the message is transmitted from the host to the message processor to which it is attached. Once the message is entirely received, the message processor examines its header, and accordingly decides on the next outgoing channel on which to transmit it. If this selected channel is busy, the message waits in a queue until the channel becomes free, at which time transmission begins.

4. A variation of message switching is packet switching. Here the message is broken up into several pieces of a given maximum length, called packets. As with message switching, each packet contains a header and a checksum. Packets are transmitted independently in a store-and-forward manner and packet switching achieves the benefits provided by message switching and offers added features.

5. With circuit switching, there is always an initial connection cost incurred in setting up the circuit. It is cost-effective only in those situations where once the circuit is set up there is a guaranteed steady flow of information transfer to amortize the initial cost. This is certainly the case with voice communication in the traditional way, and indeed circuit switching is the technique used in the telephone system.

6. Communication among computers, however, is characterized as bursty. Burstiness is a result of the high degree of randomness encountered in the message-generation process and the message size, and of the low delay constraint required by the

user. The users and devices require the communication resources relatively infrequently; but when they do, they require a relatively rapid response.

7. Therefore, for bursty users, store-and-forward transmission techniques offer a more cost-effective solution, since a message occupies a particular communications link only for the duration of its transmission on that link; the rest of the time it is stored at some intermediate message switch and the link is available for other transmissions. Thus the main advantage of store-and-forward transmission over circuit switching is that the communication bandwidth is dynamically allocated, and the allocation is done on the fine basis of a particular link in the network and a particular message.

8. Packet switching achieves the benefits discussed so far and offers added features. It provides the full advantage of the dynamic allocation of the bandwidth, even when messages are long. Indeed, with packet switching, many packets of the same message may be in transmission simultaneously over consecutive links of a path from source to destination, thus achieving a "pipelining" effects and reducing considerably the overall transmission delay of the message as compared to message switching.

9. Packet switching achieves the benefits discussed so far and offers added features. It tends to require smaller storage allocation at the intermediate switches. It also has better error characteristics and leads to more efficient error recovery procedures, as it deals with smaller entities. Needless to say, packet switching presents design problems of its own, such as the need to reorder packets of a given message that may arrive at the destination node out of sequence.

参考译文

电路交换与分组交换

交换技术有两种基本类型:电路交换与报文交换。在电路交换中,当呼叫发生时,由呼叫源点到终点之间要建立整个通路的连线,而且在通信双方释放该电路之前,此通路一直保持分配给这对源点-终点(不管通路是否使用)。被称为电路交换机(或以电话行业用语称之为局交换)的交换设备没有存储或控制用户送往终点路由的数据的能力。电路由特殊的信令建立,该信令通过网络选择路由,并在其进程中确定信道。一旦路由建立,一个返回信号就通知呼叫源开始传输(数据)。接着,由源点到终点间进行直接的数据传输,其间对该通信子网不会做任何干预。

在报文交换中,传输单元是一个被精心定义的数据块,该数据块称为报文。除了要发送的内容外,报文还包括有报头和校验项。报头含有源地址和目的地地址的信息,以及其他的控制信息,而校验项用于误码控制。交换单元是一台被称为报文处理器的计算机,它具有处理和存储的能力。报文独立并异步地传输,在源点与终点间选择自己的传送路由。首先报文由主机送往与之相连的报文处理机。一旦报文被完全收到,报文处理机就检查其报头,并相应地决定该报文传送的下一个输出信道。如果这个所选信道忙,则该报文就排队等待,直到此信道空闲再开始发送。在下一个报文处理器中,报文被再次收到、存储、检查,并在某个输出信道上再发送出去。相同的过程继续进行着,直到报文交到目的地为止。这种传送技术亦被称为存储转发传输技术。

分组交换是报文交换的一种变形。在分组交换中,报文被以指定的最大长度分成若干个被称为分组的段节。与报文交换一样,每个分组都含有一个报头和校验项。分组以存储转发方式独立传送。

在电路交换的情况下,建立电路总要对开始的接续付出代价。只有在这种情况下,即一旦电路建立后,信息的传送确保持续,源源不断,以便分摊初始花费,才能提高价格效率比。传统方式的语音通信就属于这种情况,因而电路交换技术才为电话系统所采用。但是,计算机的通信具有突发特性。突发性是由报文产生过程和报文长度的高度随机性所造成的,也是因用户对时延要求很短而造成的。用户及设备不常用到通信资源,但是当他们用到时,他们就要求其具有相当迅速的反应。如果要建立一个固定的专用的端到端电路以连接两个用户,则必须对该电路分配足够的传输带宽以合乎对时延的要求,其结果是电路的利用率很低。如果对每个报文传输要求都要建立和释放大带宽的电路,则与报文传输的时间相比,电路建立的时值将很大,造成很低的电路利用率。因此,对突发性用户(峰值速率与平均速率很高为该用户的特征),存储转发传输技术提供了一个更低价高效的解决办法,因为只有在报文传送的时间里,报文才占据一条特定的链路,其他时间,报文被存储在某个中间交换机中,因而此时的链路可用于其他传输。这样,与电路交换相比,存储转发方式的主要优点是通信带宽的动态分配,而且这种分配是以网络中的特定链路和特定报文(对一个特定的源点-终点对来说)为基础的。

分组交换除具有以上讨论的优点外,还具有一些特点。它提供带宽动态分配的全部优势,甚至当报文很长时依然如此。由于有分组交换,相同报文的多个分组可以通过源点到终点通路中的多条链路同时传送,因而达到"管道传送"的效应。与报文交换相比,它大大地减少了报文整体的传送时延。在中间交换设备中,这种方式只需要较小的存储分配区域。分组交换的误码特性较好,由于它只涉及很短的长度,因而导致了更高效的纠错方式。当然,分组交换亦有它自己设计上的麻烦,例如当报文无序地到达目的节点时,需要重新对该报文进行分组排序。

UNIT 13

PASSAGE

EPON[1]

Nowadays, since more and more Internet businesses have gradually entered into numerous households with the rapid development of the Internet, bottlenecks of access network are emerging.[2] EPON technology, which combines a mature Ethernet technology and high-bandwidth PON technology, is an ideal access method to achieve integrated services.[3]

In recent years, the telecommunication backbone network has been upgraded from time to time while more fibers are laid and devices of larger capacities have come into use. But as for the access network, the copper line is still the leading choice. The tremendous increase in Internet services has pricked up the shortage of the access network's capacity. The access network which is called "last mile" still remains the bottleneck between high-speed LANs and the high-capacity backbone network.

The most widely deployed "broadband" solutions today are DSL and CM networks. Although they are improvements compared to 56 kbit/s dial-up lines, they are still unable to provide enough bandwidth for emerging services such as video-on-demand (VOD), interactive gaming, or two-way video conferencing.[4] A new kind of access technology is required for the time with following features: inexpensive, simple to upgrade, and being able to provide bundled voice, data and video services. EPON, which merges the low-cost Ethernet technology and low-cost optic network architecture, is the best representative of the future-oriented next generation access network technologies.[5]

Ethernet passive optical networks (EPON) are next-generation subscriber access network technologies that deploy optical access lines between an optical line terminator (OLT) and multiple optical network units (ONU) and/or optical network terminators (ONT).[6] OLT is located in network side and ONU/ONT is in customer side. EPON will be utilized in providing subscribers with access bandwidths of 10 Mbit/s up to 10 Gbit/s per user to be needed for high-quality Internet services.

A typical EPON system is composed of OLT, ONU, and ODN (see Fig. 13-1).

The OLT(Optical Line Terminal) resides in the Central Office (CO) and connects the optical network to the metropolitan-area network or wide-area network, also known as the backbone or long-haul network.

OLT is both a switch or router and a multi-service platform which provides EPON-

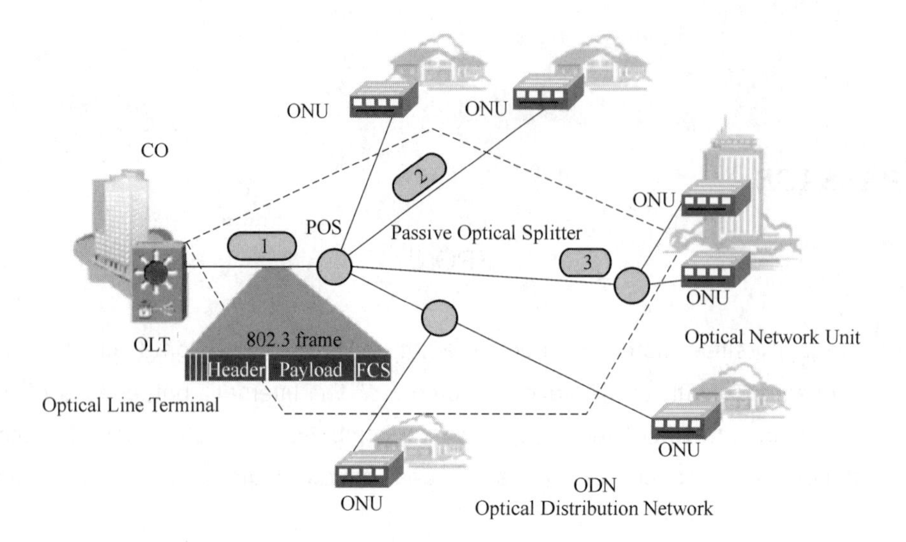

Fig. 13-1 EPON Network Architecture

oriented optical interfaces. Besides the network assembling and access functions, OLT could also perform bandwidth assignments, network security and management configurations according to the customers' different QoS requirements.

The ONU(Optical Network Unit)is located either at the end-user location or at the curb and provides optical interfaces which are connected to the OLT and service interfaces at users' side such as voice, data and video.[7]

The ODN(Optical Distributed Network)is an optical distribution network and is mainly composed of one or more passive optical splitters which connect the OLT and ONU. Its function is to split downstream signal from one fiber into several fibers and combine optical upstream signals from multiple fibers into one. Optical splitter is a simple device which needs no power and could work in an all-weather environment. The typical splitters have a splitting ratio of 2, 4, 8, 16 or 32 and can be connected to each other. The longest distance the ODN could cover is 20 km.

In the downstream direction (from OLT to ONUs), a PON is a point-to-multipoint network. In the upstream direction, a PON is a multipoint-to-point network: multiple ONUs transmit all towards one OLT. Directional properties of a passive splitter/combiner are such that an ONU's transmission cannot be detected by other ONUs.

EPON system adopts the broadcast method in downstream direction. Ethernet frames of different ONUs are transmitted in one downstream timeslot. The packet which carries all the ONUs' data reaches different ONUs and each ONU only extracts its own data frames and transmit them to the users while others frames are discarded.

However, data streams from different ONUs transmitted simultaneously still may collide. Thus, in the upstream direction (from user to network), PON should employ some channel separation mechanism to avoid[8] data collisions and fairly share the trunk fiber channel capacity and resources.

One possible way of separating the ONU's upstream channels is to use a wavelength-division multiplexing (WDM), in which each ONU operates on a different wavelength. While it is a simple solution (from a theoretical perspective), it remains cost-prohibitive for an access network.

In a time division multiplexing (TDM) PON, simultaneous transmissions from several ONUs will collide when reaching the combiner. In order to[9] prevent data collisions, each ONU must transmit in its own transmission window (timeslot). One of the major advantages of a TDM PON is that all ONUs can operate on the same wavelength. The OLT will also need a single receiver. A transceiver in an ONU must operate at the full line rate, even though the bandwidth available to the ONU is lower. However, this property also allows the TDM PON to efficiently change the bandwidth allocated to each ONU by changing the assigned timeslot size, or even employ statistical multiplexing to fully utilize the bandwidth available in the PON.[10]

It is possible that EPONs and point-to-point optical Ethernet offer the best possibility for a turnaround in the telecom sector.[11] As service providers invest in optical access technologies, this will enable new applications, stimulating revenue growth and driving more traffic onto backbone routes.

NEW WORDS AND PHRASES

gradually	*ad*. 逐步地，渐渐地；按部就班地
numerous	*a*. 很多的，许多的；数不清的
emerge	*vi*. 出现，浮现；暴露；摆脱
mature	*a*. 成熟的；仔细考虑过的
access	*n*. 接近，进入
integrate	*vt*. 使一体化；使整合；使完整
upgrade	*vt*. 提升；使升级
capacity	*n*. 容量；才能；性能
tremendous	*a*. 极大的，巨大的；可怕的，惊人的
prick	*v*. 刺；竖起，加剧
bottleneck	*n*. 瓶颈；瓶颈路段
deploy	*v*. 使展开；施展；利用
solution	*n*. 解答；解决(办法)；解释
dial-up	拨号
interactive	*a*. 互相作用的；交互式的；互动的

conference	*n*. 会议;讨论;(正式)讨论会
bundle	*v*. 收集,归拢,把……塞入
architecture	*n*. 体系结构;(总体、层次)结构
generation	*n*. 一代人,代,时代
charter	*vt*. 发给……许可证;发给特许执照
subscriber	*n*. 用户,订户
institute	*vt*. 建立;制定;开始;着手
passive	*a*. 被动的;消极的
terminator	*n*. 终结者,终结器
metropolitan	*a*. 大都会的;大城市的
router	*n*. 路由器
platform	*n*. 台;站台;平台
interface	*n*. 界面;接口;交界面
assemble	*vt*. 装配,组合
assignment	*n*. 分给,分配;任务,工作
security	*n*. 安全;保证,担保
configuration	*n*. 组合,布置;结构,构造
distribution	*n*. 分配,分布
splitter	*n*. 分裂机;分流器
upstream	*a*. 向上游的;逆流而上的
environment	*n*. 环境,外界;周围
property	*n*. 特性,属性;财产
timeslot	*n*. 时隙
packet	*n*. 小包;信息包,分组
extract	*vt*. 提取;拔出;选取
discard	*vt*. 丢弃,抛弃;解雇
collide	*vi*. 相撞;碰撞;冲突
separation	*n*. 分离,分开;间隔,距离
mechanism	*n*. 机制,机能;结构,机械装置
simultaneously	*ad*. 同时地;一齐
cost-prohibitive	成本高昂的
prevent	*vt*. 预防;阻碍,阻止
wavelength	*n*. 波长;波段
allocate	*vt*. 分配,分派;把……拨给
statistical	*a*. 统计的;统计学的

utilize　　vt. 利用,使用

NOTES

1. 本篇课文涉及接入网领域,题目为:以太无源光网络技术(EPON)。

2. since 引出原因状语从句,主句为后面的 bottlenecks of access network are emerging;with the rapid development of the Internet,介词短语,表示伴随情况。本句可译成:今天,随着互联网的快速发展,越来越多的互联网业务逐步走进千家万户,出现了接入网瓶颈。

3. which 引导非限定性定语从句,用来说明前面的 technology;to achieve integrated services 不定式短语做后置定语,修饰 method。本句可译成:EPON 技术,结合了成熟的以太网技术和高带宽 PON 技术,是获得综合业务的理想接入方式。

4. such as,比如;although 和 but 在英语中不能同时出现,但译成汉语时要译成"虽然……但是……"。本句可译成:虽然它们相比于 56 kbit/s 拨号线有了很大的改进,但仍不能为新出现的业务(如视频点播、交互游戏或双向视频会议等)提供足够的带宽。

5. which 引导非限定性定语从句,修饰 EPON;future-oriented,面向未来的。本句可译成:EPON 技术,结合了低费用以太网技术和低费用光网络技术,是面向未来的下一代接入网技术的最佳代表。

6. that 引导定语从句,修饰 technologies。本句可译成:以太无源光网络(EPON)是下一代用户接入网络技术,它在光线路终端(OLT)和许多光网络单元(ONU)或光网络终端(ONT)之间铺设光纤接入线。

7. 此句主语有两个谓语;either…or…,二者择一的;which 引导定语从句,修饰前面的 interfaces。本句可译成:ONU(光网络单元)位于终端用户的位置或路边,可以提供连接到 OLT 的光接口以及用户端如话音、数据和视频的业务接口。

8. to avoid 不定式表示目的。本句可译成:这样,在上行方向(从用户到网络),PON 应该采用一些信道分离机制以避免数据冲突并公平地共享主干光纤信道容量和资源。

9. in order to,为了,以便。本句可译成:为避免冲突,每个 ONU 必须在其自己的传输窗口(时隙)进行传输。

10. allocated to each ONU 是过去分词短语做后置定语,修饰前面的 bandwidth;changing 为动名词,做介词 by 的宾语;to fully utilize the bandwidth 不定式短语表示目的。本句可译成:然而,这一特性也允许时分复用 PON 通过改变分配的时隙大小来有效地改变每个 ONU 的分配带宽,甚至采用统计复用来完全利用 PON 的可用带宽。

11. 在这里"it"是形式主语,that 引导的从句是真正的主语。本句可译成:很有可能 EPON 和点到点光以太网为电信部门的转向提供最佳的可能性。随着业务供应商投资于光接入技术,新应用将成为可能,从而刺激收入增长并驱动骨干网上更多的业务量。

EXERCISES

一、请将下述词组译成英文

1. 成熟的以太网技术　　2. 理想的接入方法　　3. 骨干网

4. 铜线	5. 拨号线	6. 提供足够的带宽
7. 视频会议	8. 下一代接入网	9. 用户接入网技术
10. 光线路终端	11. 光网络单元	12. 多业务平台
13. 完成带宽分配	14. 采用广播方法	15. 提取自己的数据帧
16. 分离 ONU 上行信道	17. 光分配器	18. 提供光接口
19. 为防止数据冲突	20. 主要的优势	21. 安排的时隙

二、请将下述词组译成中文

1. development of the Internet
2. bottlenecks of access network
3. larger capacities
4. high-speed LANs
5. to provide enough bandwidth
6. interactive gaming
7. data and video services
8. subscriber access network technologies
9. access bandwidths of 10 Mbit/s
10. high-quality Internet services
11. the metropolitan-area network
12. multi-service platform
13. optical distribution network
14. point-to-multipoint network
15. data streams from different ONUs
16. to avoid data collisions
17. to use a wavelength-division multiplexing

三、选择合适的答案填空

1. EPON technology, which combines a mature Ethernet technology and high-bandwidth PON technology, is an ideal access method ________ integrated services.

A. to achieve　　B. achieving

C. achieve　　D. achieved

2. The access network which is ________ "last mile" still remains the bottleneck between high-speed LANs and the high-capacity backbone network.

A. calling　　B. to call

C. called　　D. call

3. EPON will be utilized in ________ subscribers with access bandwidths of 10 Mbit/s up to 10 Gbit/s per user to be needed for high-quality Internet services.

A. provide　　B. to provide

C. provided　　D. providing

4. The OLT (Optical Line Terminal) resides in the Central Office (CO) and

connects the optical network to the metropolitan-area network or wide-area network, also ________ as the backbone or long-haul network.

A. know B. known

C. knowing D. to know

5. The ODN(Optical Distributed Network)is an optical distribution network and is mainly ________ of one or more passive optical splitters which connect the OLT and ONU.

A. composed B. to compose

C. composing D. compose

6. One possible way of ________ the ONU's upstream channels is to use a wavelength-division multiplexing (WDM), in which each ONU operates on a different wavelength.

A. separate B. separated

C. separating D. to separate

7. One possible way of separating the ONU's upstream channels is ________ a wavelength-division multiplexing (WDM), in which each ONU operates on a different wavelength.

A. using B. use

C. used D. to use

四、根据课文内容选择正确答案

1. Nowadays, since more and more Internet businesses have gradually entered into numerous households ________, bottlenecks of access network are emerging.

A. with the rapid development of the Internet

B. with the rapid development of the access network

C. with the rapid development of the optical fiber technologies

D. with the rapid development of the LAN

2. The access network which is called "last mile" still remains the bottleneck between high-speed LANs and ________.

A. the high-capacity long-haul network

B. the high-capacity optical network

C. the high-capacity network

D. the high-capacity backbone network

3. EPON, which merges the low-cost Ethernet technology and ________, is the best representative of the future-oriented next generation access network technologies.

A. low-cost optic network unit

B. low-cost optic network architecture

C. low-cost optic network technology

D. low-cost optic cable modem network

4. OLT is both a switch or router and a multi-service platform which ________.

A. provides ONU-oriented optical interfaces

B. provides PON-oriented optical interfaces

C. provides EPON-oriented optical interfaces

D. provides ODN-oriented optical interfaces

5. The packet which carries all the ONUs' data reaches different ONUs and each ONU only ________ and transmit them to the users while others frames are discarded.

A. provide its own data stream

B. provide its own data frames

C. extracts its own data stream

D. extracts its own data frames

6. One possible way of ________ is to use a wavelength-division multiplexing (WDM), in which each ONU operates on a different wavelength.

A. combining the OLT's upstream channels

B. combining the ONU's upstream channels

C. separating the ONU's upstream channels

D. separating the OLT's upstream channels

7. However, this property also allows the TDM PON to efficiently ________ by changing the assigned timeslot size, or even employ statistical multiplexing to fully utilize the bandwidth available in the PON.

A. change the bandwidth allocated to each ONU

B. change the bandwidth allocated to each OLT

C. change the architecture allocated to each ONU

D. change the architecture allocated to each OLT

五、请将下述短文译成中文

1. Nowadays, since more and more Internet businesses have gradually entered into numerous households with the rapid development of the Internet, bottlenecks of access network are emerging. EPON technology, which combines a mature Ethernet technology and high-bandwidth PON technology, is an ideal access method to achieve integrated services.

2. The tremendous increase in Internet services has pricked up the shortage of the access network's capacity. The access network which is called "last mile" still remains the bottleneck between high-speed LANs and the high-capacity backbone network.

3. A new kind of access technology is required for the time with following features: inexpensive, simple to upgrade, upgradeable, and being able to provide bundled voice, data and video services. EPON, which merges the low-cost Ethernet technology and low-cost optic network architecture, is the best representative of the future-oriented next generation access network technologies.

4. Ethernet passive optical networks (EPON) are next-generation subscriber access network technologies that deploy optical access lines between an optical line terminator (OLT) and multiple optical network units (ONU) and/or optical network terminators (ONT). OLT is located in network side and ONU/ONT is in customer side.

5. Besides the network assembling and access functions, OLT could also perform bandwidth assignments, network security and management configurations according to the customers' different QoS requirements. The ONU(Optical Network Unit)is located either at the end-user location or at the curb and provides optical interfaces which are connected to the OLT and service interfaces at users' side such as voice, data and video.

6. However, data streams from different ONUs transmitted simultaneously still may collide. Thus, in the upstream direction (from user to network), PON should employ some channel separation mechanism to avoid data collisions and fairly share the trunk fiber channel capacity and resources.

7. In a time division multiplexing (TDM) PON, simultaneous transmissions from several ONUs will collide when reaching the combiner. In order to prevent data collisions, each ONU must transmit in its own transmission window (timeslot). One of the major advantages of a TDM PON is that all ONUs can operate on the same wavelength.

8. It is possible that EPONs and point-to-point optical Ethernet offer the best possibility for a turnaround in the telecom sector. As service providers invest in optical access technologies, this will enable new applications, stimulating revenue growth and driving more traffic onto backbone routes.

参考译文

以太无源光网络技术(EPON)

今天，随着互联网的快速发展，越来越多的互联网业务逐步走进千家万户，接入的瓶颈效应开始浮现。EPON 技术，结合了成熟的以太网技术和高带宽 PON 技术，是综合业务理想的接入方式。

近年来，电信骨干网不断更新，铺设了更多的光纤，更大容量的电信设备投入了使用。但是，但对于接入网而言，铜线接入仍为首选。互联网业务的剧增暴露了接入网容量的不足。被称为“最后一公里”的接入网仍旧是高速局域网和大容量骨干网间的瓶颈。

今天广泛采用的宽带解决方案是 DSL 和 CM 网络。虽然它们相比于56 kbit/s拨号接入来说有了很大的改进，但仍不能为新出现的业务(如视频点播、交互游戏或双向视频会议等)提供足够的带宽。此时需要一种新型的接入技术，它应有如下特性：便宜、升级简单并同时提供声音、数据和视频业务。EPON 技术，结合了低费用以太网技术和低费用光网络技术，是面向未来的下一代接入网技术的最佳代表。

以太无源光网络(EPON)是下一代用户接入网络技术，它在光线路终端(OLT)和许多光网络单元(ONU)或光网络终端(ONT)之间铺设光纤接入线。光线路终端(OLT)位于网络侧,而光网络单元或光网络终端(ONU/ONT)在用户侧。需要高质量互联网业务的用户可以使用EPON来提供10 Mbit/s至10 Gbit/s的接入带宽。典型的EPON系统由OLT、ONU和ODN组成(见图13-1,第154页)。

OLT(光线路终端)位于中心局(CO),它将光网络连到城域网或广域网(也称为骨干网或长途网)。OLT既是交换机或路由器也是提供面向EPON光接口的多业务平台。除了网络汇集和接入功能外,OLT也可以根据用户的不同服务质量需求完成带宽分配、网络安全和管理配置。

ONU(光网络单元)位于终端用户的位置或路边,可以提供连接到OLT的光接口以及用户端如话音、数据和视频的业务接口。

ODN(光分配网)是光纤分配网络,主要由一个或多个连接OLT和ONU的无源光分离器组成。其功能是从一根光纤分离下行信号到多个光纤和从多个光纤合并上行光信号到一根光纤。光分配器是不需要电源的简单器件,可以工作在全天候环境下。典型的分配器具有2、4、8、16或32的分配率,并且可以彼此连接。ODN可以覆盖的最长距离是20千米。

在下行方向(从OLT到多个ONU),PON是点到多点网络结构。在上行方向,PON是多点到点结构:多个OUN都向OLT传信号。无源分配器/组合器的方向特性使得一个ONU的传输信号不会被其他ONU检测到。

EPON系统在下行方向采用广播方式。不同ONU的以太帧在一个下行时隙中传输。携带所有ONU数据的分组到达不同的ONU,每个ONU只提取自己的数据帧并将数据帧传到用户,而其他的数据帧被丢弃。

然而,从不同ONU同时传输的数据流仍会产生冲突。这样,在上行方向(从用户到网络),PON应该采用一些信道分离机制以避免数据冲突并公平地共享主干光纤信道容量和资源。

分离ONU上行信道的一种可能方式是使用波分复用(WDM)技术,在这种技术中,每个ONU都工作在不同的波长。虽然这是一种简单的解决方案(从理论角度),但对于接入网来说,费用很高。

在时分复用(TDM)PON技术中,当从几个ONU同时传输的数据到达组合器时会产生冲突。为避免冲突,每个ONU必须在其自己的传输窗口(时隙)进行传输。TDM PON的主要优点是所有的ONU都工作在同样的波长。OLT需要单个的接收机。即使ONU的可用带宽较低,ONU中的收发信机也必须工作在满线路速率上。然而,这一特性也允许时分复用PON通过改变分配的时隙大小来有效地改变每个ONU的分配带宽,甚至采用统计复用来完全利用PON的可用带宽。

很有可能,EPON和点到点光以太网为电信部门的转向提供最佳的可能性。随着业务供应商投资于光接入技术,新应用将成为可能,从而刺激收入增长并驱动骨干网上更多的业务量。

UNIT 14

PASSAGE

IPTV[1]

Over the past few months, there has been increased interest in the use of IP networks to deliver broadcast-quality TV. IPTV is a relatively recent buzzword having emerged only three or four years ago.[2] In a nutshell, it means the delivery of television over Internet protocol (IP) networks. But for the telecommunication industry, which drives it to the market, IPTV means much more. For them, IPTV is an entirely new multimedia experience extending the borders of conventional broadcast television; it is an integrated, "all-embracing" media platform offering a bundle of diverse content and communication services from a single provider over a single network to a single user device — all with a single payment.[3]

Why use 30-year-old technology designed for slow, unreliable, unicast communications to transport video? The answer is that IP is a magical platform to breed success and it has become totally pervasive.[4] The second motivation is that IP can offer more functionality than traditional TV. The open, extensible nature of IP holds the promise of many more services in the future. IP infrastructure also can be leveraged to provide video telephony and remote monitoring, as well as new uses of video within applications that haven't been thought of yet.[5] The final factor that is emerging is cost savings.

There are diverging views about what IPTV really means.[6] To broadcasters, IPTV is simply a new emerging platform for distributing digital television channels to home consumers using a TV screen. To the telecom industry, IPTV is synonymous with a new broadband digital technology, offering voice, data and video. IPTV is complementary to existing satellite, cable and terrestrial systems, although in some cases it may become a vigorous competitor to them.

So far, the telecom industry has merely been providing telecommunication services such as voice connections between two points.[7] The telecom companies were not at all concerned with the content of the information carried. They are now getting involved in IPTV as they facing decreasing subscriber revenues from their traditional voice and broadband communication.[8] They are in the process of moving into the content-providing domain with no real expertise and experience in providing television services to the general public, simply in order to improve their balance sheet and bring the customers back. Indeed, IPTV is beginning to look like a promising new business opportunity for telecom operators.

However, telecoms will have to resolve some major technical challenges. One problem is the end-to-end transmission quality due to limited network bandwidth which decreases with the distance from the exchange cabinet. New modulation strategies such as ADSL2+ and VDSL are being introduced, but the cost concerned is relatively high. The introduction of high-definition television (HDTV) may be quite a challenge, even if advanced coding technologies such as ITU/MPEG 4 or Microsoft VC1 are used.[9] Furthermore, telecom networks do not generally serve the hundreds of thousands and even millions of simultaneous users. A lack of international standardization, which is currently the main obstacle to establishing horizontal markets and equipment inter-operability, is now being addressed by the International Telecommunications Union.[10]

IPTV can potentially offer a myriad of new innovative services and applications to the user, many of which are already possible with digital television, the main differences come from the fact that IPTV uses a two-way communication channel, so the user can interact directly with the content and service provider.[11] The interactive link between the provider and the user enables sending individual video streams to individual devices in the home at the user's request, in contrast to the broadcast model where all channels are sent to all users all the time. Such personalized services are becoming increasingly popular, particularly with entertainment and education.

While the technical issues are not easy, they are not insurmountable. The legal and regulatory issues, however, are a real nightmare. One of the outstanding issues is to determine which national regulatory body is responsible for the regulation of IPTV services. This depends on the very definition of IPTV. Some countries consider IPTV a broadcast service and some a telecom service. Not only do the legal regimes vary from country to country, they are also different for telecoms and broadcasters.[12]

IPTV is still in its infancy. Its market size is still small but it is growing rapidly. The prospects for the commercial success are great. The principle drivers for IPTV are the incumbent telecom and Internet companies. So far, broadcasters have been marginally involved in the IPTV process. It is now time to raise the awareness of broadcasters, so that they can play a more active role in the IPTV developments.[13]

NEW WORDS AND PHRASES

buzzword　　*n*. 时髦术语,口号
emerge　　*vi*. 出现,浮现
nutshell　　*n*. 极小的东西
in a nutshell　　概括地说,一句话
market　　*n*. 市场

entirely	ad.完全地,全部地
experience	n.经历
extend	v.伸长,延伸
border	n.边界
integrate	v.综合,集成
embrace	v. 拥抱,怀抱;利用,抓住,采用;包围,包括,包含
platform	n.平台,站台,讲台
bundle	n.一捆,一束,一大堆
diverse	a.多种多样的,形形色色的
content	n.内容,含量,目录
payment	n.支付,支付的款项
unreliable	a.不可靠的
unicast	a.单播的
magical	a.迷人的,神秘的,有魔力的
breed	v.养育,培养,教养
pervasive	a.遍布的,充满的
motivation	n.动力,动机,诱因
functionality	n.功能
extensible	a.可延伸的,可扩展的
promise	n.允诺,诺言;出息,前途
infrastructure	n.基础结构
leverage	v.举债经营
diverge	v.分叉,分歧;发散
distribute	vt.分配,分发,分送;散布
synonymous	a.同义的
complementary	a.补充的,补足的
terrestrial	a.地面的
vigorous	a.朝气蓬勃的
merely	ad.仅仅,只不过
revenue	n.(年)收入,收益
domain	n.领域,领地,范围
expertise	n.专长,专门的技能
balance	n.平衡,均衡
sheet	n.表格
challenge	n.挑战

cabinet	n. 机柜
strategy	n. 战略,策略
obstacle	n. 障碍
horizontal	a. 同行业的,同一层次的
potentially	ad. 潜在地,有可能地
myriad	n. 无数
innovative	a. 创新的
interact	a. 交互的
personalize	vt. 个人化
entertainment	n. 娱乐
insurmountable	a. 难以逾越的
legal	a. 法律的,合法的,法定的
regulatory	n. 管理,控制
nightmare	n. 噩梦,梦魇
regime	n. 政治制度,社会制度,体制;政权
infancy	n. 婴幼年期,初期
prospect	n. 展望,前景,前程
incumbent	a. 成为责任的,义不容辞的
marginally	ad. 边沿,勉强够格地
awareness	n. 意识,认识

NOTES

1. 本篇课文涉及网络电视的有关内容,题目可译为:IP 电视。

2. having emerged,现在分词的完成时态,修饰 buzzword。本句可译成:IPTV 是近来较为流行的词语,其出现也只是在三四年之前。

3. extending the borders of conventional broadcast television,现在分词短语,修饰 experience。offering a bundle of diverse content and communication services,现在分词短语,修饰 platform。本句可译成:IPTV 是一种全新的多媒体体验,它扩展了传统广播电视的范围;IPTV 是一种综合的、"包含一切的"媒体平台,可以从单一业务供应商通过单一网络向单一用户设备提供多种服务内容和通信服务——所有这些只需一笔支付。

4. that IP is a magical platform to breed success,表语从句,做 that 前面 is 的表语。本句可译成:答案是 IP 是孕育成功的神奇平台,IP 已经无孔不入。

5. 此句为被动句。to provide video telephony and remote monitoring,不定式短语表示目的。that haven't been thought of yet,定语从句,修饰 application。本句可译成:IP 的体系结构也可以用来提供视频电话和远程监控,以及一些人们还没有想到的新的视频应用。

6. what IPTV really means,由 what 引导的宾语从句,做 about 的宾语。本句可译成:关于 IPTV 的真实含义有多种观点。

7. so far,至此,迄今为止。本句用的是完成进行时态。such as,例如。本句可译成:迄今为止,电信行业一直只是提供如两点间话音通信的电信业务。

8. as they facing decreasing subscriber revenues,由 as 引出时间状语从句。本句可译成:当电信公司面临其传统话音和宽带通信日益减少的用户收入时,他们开始涉足 IPTV。

9. even if,即使。such as,如。本句可译成:引入高清晰电视可能具有相当的挑战,即使采用像 ITU/MPEG 4 或微软的 VC1 这样的先进编码技术。

10. which is currently the main obstacle to establishing horizontal markets and equipment inter-operability,非限定性定语从句,修饰前面部分。本句可译成:缺少国际标准是目前建立统一市场和设备互通的主要障碍,这一问题也正为国际电信联盟所关注。

11. that IPTV uses a two-way communication channel,同位语从句,说明 fact。so 是连接词,表示"因而"。本句可译成:主要的区别在于 IPTV 使用双向通信通道,所以用户可以与业务内容和业务供应商直接进行交互。

12. not only…also…,不但……而且……。本句可译成:不但是国家的司法体制不同,而且其电信公司和广播业务服务商也不同。

13. so that,以便。play a more active role,起更积极的作用。本句可译成:该是提醒广播业务服务商的时候了,以便其能够在 IPTV 的发展过程中发挥积极的作用。

EXERCISES

一、请将下述词组译成英文

1. 通信行业　2. 广播电视　3. 基础结构　4. 可视电话
5. 宽带数字技术　6. 端到端的传输　7. 交换机架　8. 高清电视
9. 编码技术　10. 国际电信联盟　11. 内容提供商　12. 交互式链路

二、请将下述词组译成中文

1. the delivery of television over Internet Protocol (IP) networks
2. the diverse content and communication services
3. the video telephony and remote monitoring
4. to distribute digital television channels to consumers using a TV screen
5. the telecommunication services such as voice connections
6. the limited network bandwidth
7. the advanced coding technologies
8. the International Telecommunications Union
9. a myriad of new innovative services and applications to the user

三、选择合适的答案填空

1. Over the past few months, there has been increased interest in the use of IP networks ________ broadcast-quality TV.

A. delivered B. to deliver

C. to be delivered D. be delivered

2. In a nutshell, IPTV means the delivery of television ________ Internet Protocol (IP) networks.

A. in B. on

C. with D. over

3. IPTV is an "all-embracing" media platform ________ a bundle of diverse content and communication services from a single provider over a single network to a single user device.

A. provided B. offer

C. provide D. offering

4. To broadcasters, IPTV is simply a new emerging platform for distributing digital television channels to home consumers ________ a TV screen.

A. using B. usage

C. utilize D. use

5. To the telecom industry, IPTV is synonymous with a new broadband digital technology, ________ voice, data and video.

A. offer B. offering

C. to offer D. provider

四、根据课文内容选择正确答案

1. The telecom companies are involved in IPTV as they facing ________ from their traditional voice and broadband communication.

A. increased revenues B. more subscribers

C. decreasing revenues D. the request of the broadcasters

2. The IPTV uses ________, so the user can interact directly with the content and service provider.

A. a two-way communication channel

B. single communication channel

C. radio channel

D. microwave communication channel

3. The interactive link between the provider and the user enables sending individual video streams to individual devices in the home ________.

A. all the time B. at the user's request

C. at 6 o'clock D. in the evening

4. IPTV is a platform offering a bundle of diverse content and communication services from a single provider over ________ to a single user device.

A. a single network B. a telephone network

C. a mobile network D. optical fiber

5. To the telecom industry, IPTV is synonymous with a new broadband digital technology, offering ________. IPTV is complementary to existing satellite, cable and terrestrial systems.

A. narrow ISDN services　　B. voice conversations

C. speech channels　　D. voice, data and video

五、请将下述短文译成中文

1. In a nutshell, IPTV means the delivery of television over Internet Protocol (IP) networks. For telecoms, IPTV is an entirely new multimedia experience extending the borders of conventional broadcast television; it is an integrated, "all-embracing" media platform offering a bundle of diverse content and communication services from a single provider over a single network to a single user device.

2. Why use IP technology to transport video? The answer is that IP is a magical platform to breed success and it has become totally pervasive. The second motivation is that IP can offer more functionality than traditional TV. The open, extensible nature of IP holds the promise of many more services in the future. IP infrastructure also can be leveraged to provide video telephony and remote monitoring, as well as new uses of video.

3. There are diverging views about what IPTV really means. To broadcasters, IPTV is simply a new emerging platform for distributing digital television channels to home consumers using a TV screen. To the telecom industry, IPTV is synonymous with a new broadband digital technology, offering voice, data and video. IPTV is complementary to existing satellite, cable and terrestrial systems, although in some cases it may become a vigorous competitor to them.

4. So far, the telecom industry has merely been providing telecommunication services such as voice connections between two points. The telecom companies were not at all concerned with the content of the information carried. They are now getting involved in IPTV as they facing decreasing subscriber revenues from their traditional voice and broadband communication.

5. IPTV can potentially offer a myriad of new innovative services and applications to the user, many of which are already possible with digital television. The main differences come from the fact that IPTV uses a two-way communication channel, so the user can interact directly with the content and service provider. The interactive link between the provider and the user enables sending individual video streams to individual devices in the home at the user's request, in contrast to the broadcast model where all channels are sent to all users all the time.

6. While the technical issues are not easy, they are not insurmountable. The legal and regulatory issues, however, are a real nightmare. One of the outstanding issues is to determine which national regulatory body is responsible for the regulation of IPTV

services. This depends on the very definition of IPTV. Some countries consider IPTV a broadcast service and some a telecom service. Not only do the legal regimes vary from country to country, they are also different for telecoms and broadcasters.

参考译文

IP 电视

在过去的几个月,人们对使用 IP 网络来提供广播质量的电视节目很感兴趣。IPTV 是近来较为流行的词语,其出现也只是在三四年之前。简单来说,IPTV 意味着通过 IP 网络提供电视节目。但是,对于推动 IPTV 进入市场的电信公司来说,IPTV 的意义远不止这些。IPTV 是一种全新的多媒体体验,它扩展了传统广播电视的范围;IPTV 是一种综合的、“包罗万象的”媒体平台,可以从单一业务供应商通过单一网络向单一用户设备提供多种服务内容和通信服务——所有这些只需单一支付。

为什么使用已存在 30 年的为低速、不可靠和单播通信设计的技术来传输视频?答案是:IP 是孕育成功的奇妙平台,IP 已经无孔不入。第二个动机是 IP 能够提供比传统电视更多的功能。IP 的开放和扩展特性使其有希望在未来提供更多的服务。IP 的体系结构也可以用来提供视频电话和远程监控,以及一些人们还没有想到的新的视频应用。最后一个因素是节省费用。

关于 IPTV 的含义有多种观点。对于广播业务服务商来说,IPTV 仅仅是一种新出现的使用 TV 屏幕向家庭用户发布数字电视频道的平台。对于电信业来说,IPTV 与提供话音、数据和视频的新的宽带数字技术同义。IPTV 是对现有卫星、有线系统和地面系统的补充,虽然有时候是现有系统有力的竞争者。

迄今为止,电信行业一直只是提供如两点间话音通信的电信业务。电信公司并不关心所承载的信息内容。当电信公司面临其传统话音和宽带通信日益减少的用户收入时,他们开始涉足 IPTV。他们正在进入内容服务领域,在向大众提供电视服务方面并没有专业知识和经验,而只是为了改善财务平衡和抢回用户。确实,IPTV 对电信运营商来说正开始有望成为新的商业机会。

然而,电信公司必须要解决一些主要的技术问题。一个问题是由于有限的带宽所带来端到端的传输质量问题,带宽会随着离交换机距离的增加而下降。正在引入新的调制解调技术,如 ADSL2+和 VDSL,但相应的费用较贵。引入高清晰电视可能具有相当的挑战,即使采用像 ITU/MPEG 4 或微软的 VC1 这样的先进编码技术。而且,电信网络通常并不适合为成千上万的用户同时提供服务。缺少国际标准是目前建立统一市场和设备互通的主要障碍,这一问题也正为国际电信联盟所关注。

IPTV 有潜力向用户提供大量的新型业务,许多这类新业务已经在数字电视中成为可能,主要的区别在于 IPTV 使用双向通信通道,所以用户可以与业务内容和业务供应商直接进行交互。供应商和用户间的交互链路能够按用户的请求向用户家中的单一设备发送单一视频流,相比来说,广播方式是将所有信道同时发送给所有用户。这种个性化的服务正变得越来越流行,特别是在娱乐和教育业务中。

虽然技术问题并不简单，但也不是不可逾越的。然而，法律和规则问题才真正让人头疼。一个突出的问题是确定哪一个国家管理机构负责IPTV业务的管理。这取决于对IPTV的确切定义。一些国家认为IPTV是广播业务，而另一些国家认为它是电信业务。不但国家的司法体制不同，而且其电信公司和广播业务服务商也不同。

IPTV仍旧还处于发展初期。市场规模还很小，但其发展迅速。商业成功的前景是很大的。IPTV主要的驱动力来自职责所在的电信公司和互联网公司。迄今为止，广播业务服务商已经开始涉足IPTV进程。该是提醒广播业务服务商的时候了，以便其能够在IPTV的发展过程中发挥积极的作用。

UNIT 15

PASSAGE

Blockchain[1]

Blockchain is a term used in information technology.[2] In essence, it is a shared database, in which the data or information is stored, with such characteristics as "non-forgery" "trace" "traceable" "open and transparent" and "collective maintenance". Based on these characteristics, blockchain technology has laid a solid foundation of "trust", created a reliable "cooperation" mechanism, and has a broad application prospect.

A blockchain, originally block chain, is a growing list of records called blocks, which are linked using cryptography.[3] Each block contains a cryptographic hash of the previous block, a timestamp, and transaction data (generally represented as a Merkle tree). By design, a blockchain is resistant to modification of the data. It is "an open, distributed ledger that can record transactions between two parties efficiently and in a verifiable and permanent way".[4] For use as a distributed ledger, a blockchain is typically managed by a peer-to-peer network collectively adhering to a protocol for inter-node communication and validating new blocks.

Once recorded, the data in any given block cannot be altered retroactively without alteration of all subsequent blocks, which requires consensus of the network majority.[5] Although blockchain records are not unalterable, blockchains may be considered secure by design. Blockchain was invented by a person using the name Satoshi Nakamoto in 2008 to serve as the public transaction ledger of the cryptocurrency bitcoin. The identity of Satoshi Nakamoto is unknown. The invention of the blockchain for bitcoin made it the first digital currency to solve the double-spending problem without the need of a trusted authority or central server.

The bitcoin design has inspired other applications, and blockchains which are readable by the public are widely used by cryptocurrencies.[6] Blockchain is considered a type of payment. Private blockchains have been proposed for business use. Sources such as *Computerworld* called the marketing of such blockchains without a proper security model "snake oil".

The first work on a cryptographically secured chain of blocks was described in 1991 by Stuart Haber and W. Scott Stornetta. They wanted to implement a system where document timestamps could not be tampered with. In 1992, Bayer, Haber and Stornetta incorporated Merkle trees to the design, which improved its efficiency by allowing several document certificates to be collected into one block.[7] Nakamoto improved the design in an important way using a Hashcash-like method to add blocks to the chain without requiring them to be signed by a trusted party. The design was implemented in 2009 by Nakamoto as a core component of the cryptocurrency bitcoin, where it serves as the public ledger for all transactions on the network.[8]

Currently, there are three types of blockchain networks — public blockchains, private blockchains and consortium blockchains.

A public blockchain has absolutely no access restrictions. Anyone with an Internet connection can send transactions to it as well as become a validator (i. e. , participate in the execution of a consensus protocol). Some of the largest, most known public blockchains are the bitcoin blockchain and the Ethereum blockchain.

A private blockchain is permissioned. One cannot join it unless invited by the network administrators. Participant and validator access is restricted. This type of blockchains can be considered a middle-ground for companies that are interested in the blockchain technology in general but are not comfortable with a level of control offered by public networks.[9] Typically, they seek to incorporate blockchain into their accounting and record-keeping procedures without sacrificing autonomy and running the risk of exposing sensitive data to the public internet.

A consortium blockchain is often said to be semi-decentralized. It, too, is permissioned but instead of a single organization controlling it, a number of companies might each operate a node on such a network. The administrators of a consortium chain restrict users' reading rights as they see fit and only allow a limited set of trusted nodes to execute a consensus protocol.

A blockchain is a decentralized, distributed and public digital ledger that is used to record transactions across many computers so that any involved record cannot be altered retroactively, without the alteration of all subsequent blocks.[10] This allows the participants to verify and audit transactions independently and relatively inexpensively. A blockchain database is managed autonomously using a peer-to-peer network and a distributed time stamping server. The use of a blockchain removes the characteristic of infinite reproducibility from a digital asset. It confirms that each unit of value was transferred only once, solving the long-standing problem of double spending.[11] A blockchain has been described as a value-exchange protocol. A blockchain can maintain title rights because, when properly set up to detail the exchange agreement, it provides a record that compels offer and acceptance.[12]

By storing data across its peer-to-peer network, the blockchain eliminates a number of risks that come with data being held centrally. Peer-to-peer blockchain networks lack centralized points of vulnerability that computer crackers can exploit; likewise, it has no central point of failure.[13] Blockchain security methods include the use of public-key cryptography. A public key (a long, random-looking string of numbers) is an address on the blockchain. Value tokens sent across the network are recorded as belonging to that address. A private key is like a password that gives its owner access to their digital assets. Data stored on the blockchain is generally considered incorruptible.

Open blockchains are more user-friendly than some traditional ownership records, which, while open to the public, still require physical access to view.[14] Because all early blockchains were permissionless, controversy has arisen over the blockchain definition. An issue in this ongoing debate is whether a private system with verifiers tasked and authorized (permissioned) by a central authority should be considered a blockchain. Proponents of permissioned or private chains argue that the term "blockchain" may be applied to any data structure that batches data into time-stamped blocks.[15] The great advantage to an open, permissionless, or public, blockchain network is that guarding against bad actors is not required and no access control is needed.[16] This means that applications can be added to the network without the approval or trust of others, using the blockchain as a transport layer.[17]

Blockchain technology can be integrated into multiple areas. The primary use of blockchains today is as a distributed ledger for cryptocurrencies, most notably bitcoin. Most cryptocurrencies use blockchain technology to record transactions. For example, the bitcoin network and Ethereum network are both based on blockchain.

Blockchain-based smart contracts are proposed contracts that could be partially or fully executed or enforced without human interaction.[18] One of the main objectives of a smart contract is automated escrow. An IMF staff discussion reported that smart contracts based on blockchain technology might reduce moral hazards and optimize the use of contracts in general. But "no viable smart contract systems have yet emerged." Due to the lack of widespread use their legal status is unclear.

Major portions of the financial industry are implementing distributed ledgers for use in banking. Banks are interested in this technology because it has potential to speed up back office settlement systems. Some video games are based on blockchain technology. The first such game, Huntercoin, was released in February, 2016. Another blockchain game is CryptoKitties, launched in November 2017. The game made headlines in December 2017 when a cryptokitty character-an-in game virtual pet—was sold for $ 100 000. There are a number of efforts and industry organizations working to employ blockchains in supply chain logistics and supply chain management.

NEW WORDS AND PHRASES

blockchain　　*n*. 区块链

forgery	*n*. 伪造，伪造物
traceable	*a*. 可追踪的，可描绘的
transparent	*a*. 透明的，显然的
cryptography	*n*. 密码学，密码
hash	*n*. 哈希，散列
timestamp	*n*. 时间戳，时间标记
resistant	*a*. 抵抗的，反抗的
modification	*n*. 修改，修正，修饰
ledger	*n*. 总账，账簿
permanent	*a*. 永久的，永恒的
validating	*a*. 确认的，有效的
retroactively	*ad*. 追溯的，逆动的
unalterable	*a*. 不变的，不能改变的
decentralized	*a*. 分散管理的，去中心的
consensus	*n*. 一致，共识
bitcoin	*n*. 比特币
inspire	*vt*. 激发，鼓舞
implement	*vt*. 实施，执行
tamper	*v*. 做手脚，破坏
conceptualize	*v*. 使概念化
restriction	*n*. 限制，约束
validator	*n*. 验证器，验证程序
permission	*n*. 允许，许可
participant	*n*. 参与者，参加者
sacrifice	*v*. 牺牲，献祭
autonomy	*n*. 自制，自主权
verify	*vt*. 核实，查证
inexpensively	*ad*. 廉价地，不贵地
infinite	*a*. 无限的，无穷的
reproducibility	*n*. 再现性，再生性
compel	*vt*. 强迫，迫使
eliminate	*vt*. 消除，排除
vulnerability	*n*. 易损性，脆弱性
cracker	*n*. 黑客，解密高手
exploit	*vt*. 开发，利用

token	n. 代币,记号,令牌
incorruptible	a. 清廉的,不能收买的
permissionless	n. 无须许可
proponent	n. 支持者,建议者
batch	n. 一批,批处理
contract	n. 合同,契约
execute	vt. 实行,执行
escrow	n. 第三方托管,第三方支付
hazard	n. 危险,风险
optimize	v. 使最优化,使完善
logistics	n. 物流,后勤

NOTES

1. 区块链,也可以写成 The Chain of Block。本文介绍了区块链的概念及一些应用。

2. used 过去分词做后置定语,修饰 term。本句译文:区块链是一个信息技术领域的术语。

3. called blocks 过去分词短语做后置定语,修饰 records;which 引导非限定性定语从句,using crytography 现在分词短语表示伴随情况。本句译文:区块链,最初称作块链,是一个不断增长的称为块的记录列表,它使用密码进行链接。

4. it 是代词,代替前面的 blockchain;that 引导定语从句,修饰 ledger。本句译文:它是"一个开放的、分布式的账本,可以有效地、可验证地、永久地记录双方之间的交易"。

5. 本句是被动句;once recorded,一旦记录下来;which 引导非限定性定语从句。本句译文:一旦被记录下来,任何给定块中的数据都不能在不改变所有后续块的情况下进行回溯性修改,这需要网络多数人的一致同意。

6. 本句中由 and 连接两个并列句;which 引导限定性定语从句,修饰 blockchain。本句译文:比特币的设计启发了其他应用,公众可读的区块链被加密货币广泛使用。

7. which 引导非限定性定语从句。本句译文:在 1992 年,Bayer、Haber 和 Stornetta 将 Merkle 树纳入设计中,将多个文档证书收集到一个块中从而提高了其效率。

8. 本句是被动句;core component,核心组件;where 引导定语从句,其先行词是表示地点的名词 bitcoin。本句译文:该构想于 2009 年由中本聪实施,作为加密货币比特币的核心组成部分,在比特币中,它是网络上所有交易的公共总账。

9. 本句是被动句;that 引导定语从句,修饰 computers;offered by public network 过去分词短语做后置定语,修饰 a level of control。本句译文:对于那些对区块链技术感兴趣但又不适应公共网络管制水平的公司来说,这种类型的区块链可以被认为是一个中间地带。

10. that 引导定语从句,修饰 ledger;so that,以便,引导目的状语从句。本句译文:区块链是一种去中心的、分布式的、公共的数字账本,它用于记录跨许多计算机的交易,因此任何涉及的记录都不能在不更改所有后续块的情况下进行回溯性更改。

11. that 引导宾语从句，做动词 confirms 的宾语；only once，只有一次；solving the long-standing problem of double spending 为分词短语，表示伴随情况。本句译文：它证实了每价值单位只转移一次，解决了长期存在的重复支出问题。

12. 本句是现在完成时态的被动句；when 引导状语从句；set up，建立；that 引导定语从句，修饰 record。本句译文：区块链可以维护所有权，因为当正确地设置交换协议的细节时，它提供了一个记录来强制要约与承诺。

13. vulnerability，脆弱性，计算机安全隐患；that 引导定语从句，修饰 point of vulnerability；likewise，同样的，也；point of failure，故障点。本句译文：点对点区块链网络缺乏计算机黑客可以利用的集中弱点；同样，它也没有中心故障点。

14. more … than 比较级；which 引导非限定性定语从句。本句译文：开放的区块链比一些传统的所有权记录更容易使用，这些记录虽然对公众开放，但仍然需要物理访问才能查看。

15. that the term "blockchain" may be applied to any data structure that batches data into time-stamped blocks 是由 that 引导的宾语从句，做 argue 的宾语；that batches data into time-stamped blocks 是由 that 引导的同位语从句，说明 data structure 。本句译文：许可链或私有链的支持者认为，术语"区块链"可能适用于将数据批处理为时间戳块的任何数据结构。

16. that 引导表语从句，做系动词 is 的表语；guarding against，防范。本句译文：对于开放的、无须许可的或公共的区块链网络来说，最大的优点是不需要防范坏人，也不需要访问控制。

17. that 引导宾语从句，做谓语动词 mean 的宾语；using the blockchain as a transport layer 为分词短语，表示伴随情况。本句译文：这意味着使用区块链作为传输层可以将应用程序添加到网络中，而不需要其他人的批准或信任。

18. 本句是被动句；blockchain-based，基于区块链的；that 引导定语从句，修饰 contracts 。本句译文：基于区块链的智能合同是一种建议的合同，它可以部分或全部执行，也可以在没有人工交互的情况下强制执行。

EXERCISES

一、请将下述词组译成英文

1. 没有访问限制	2. 用于信息技术的术语	3. 共享数据库
4. 可靠的合作机制	5. 广泛的应用前景	6. 分布式账簿
7. 在任意给定块的数据	8. 加密货币比特币账簿	9. 限制用户阅读权限
10. 第一个数字货币	11. 文件时间戳	12. 核实和审计交易
13. 强制要约与承诺	14. 存储在区块链上的数据	15. 前一个块的密码散列
16. 一致协议的执行		

二、请将下述词组译成中文

1. without alteration of all subsequent blocks

2. to serve as the public transaction ledger
3. invited by the network administrators
4. a level of control offered by public networks
5. to record transactions across many computers
6. the long-standing problem of double spending
7. to lack centralized points of vulnerability
8. to batches data into time-stamped blocks
9. to use the blockchain as a transport layer
10. to use blockchain technology to record transactions
11. the blockchain-based smart contracts
12. the main objectives of a smart contract
13. to use a peer-to-peer network and a distributed time stamping server
14. to allow several document certificates to be collected into one block
15. to incorporate blockchain into their accounting and record-keeping procedures
16. a core component of the cryptocurrency bitcoin
17. the public ledger for all transactions on the network
18. the bitcoin blockchain and the Ethereum blockchain
19. the risk of exposing sensitive data to the public internet

三、选择合适的答案填空

1. In essence, it is a ________ database, in which the data or information is stored, with such characteristics as "non-forgery" "trace" "traceable" "open and transparent" and "collective maintenance".

A. share　　B. to share
C. shared　　D. sharing

2. For use as a distributed ledger, a blockchain is typically ________ by a peer-to-peer network collectively adhering to a protocol for inter-node communication and validating new blocks.

A. manage　　B. to manage
C. managing　　D. managed

3. The invention of the blockchain for bitcoin made it the first digital currency ________ the double-spending problem without the need of a trusted authority or central server.

A. to solve　　B. solving
C. solve　　D. solved

4. Nakamoto improved the design in an important way ________ a Hashcash-like method to add blocks to the chain without requiring them to be signed by a trusted party.

A. to use　　B. using

C. used　　D. use

5. Currently, there are three types of blockchain networks — public blockchains, private blockchains and consortium ________.

A. bitcoin　　B. blockchains

C. authority　　D. blocks

6. This allows the participants ________ and audit transactions independently and relatively inexpensively. A blockchain database is managed autonomously using a peer-to-peer network and a distributed time stamping server.

A. verify　　B. verified

C. to verify　　D. verifying

7. By ________ data across its peer-to-peer network, the blockchain eliminates a number of risks that come with data being held centrally.

A. to store　　B. store

C. stored　　D. storing

四、根据课文内容选择答案

1. Each block contains ________, a timestamp, and transaction data (generally represented as a Merkle tree).

A. a trusted authority or central server

B. a cryptographic hash of the previous block

C. a protocol for inter-node communication

D. a person (or group of people) known as Satoshi Nakamoto

2. Blockchain was invented by a person using the name Satoshi Nakamoto in 2008 to serve as ________ of the cryptocurrency bitcoin.

A. cryptographic hash of the previous block

B. a proper security model "snake oil"

C. the double-spending problem

D. the public transaction ledger

3. The first work on a cryptographically secured chain of blocks was described in 1991 by Stuart Haber and W. Scott Stornetta. They wanted to implement a system where ________.

A. Blockchain is considered a type of payment rail

B. it serves as the public ledger for all transactions

C. document timestamps could not be tampered with

D. most known public blockchains are the bitcoin blockchain

4. A private blockchain is permissioned. One cannot join it unless ________.

A. invited by the network administrators

B. interested in the blockchain technology

C. used to record transactions across many computers

D. managed autonomously using a peer-to-peer network

5. Typically, they seek to ________ and record-keeping procedures without sacrificing autonomy and running the risk of exposing sensitive data to the public internet.

A. incorporate blockchain into their accounting

B. implement a system where document timestamps

C. be signed by a trusted party

D. a protocol for inter-node communication

6. ________ or private chains argue that the term "blockchain" may be applied to any data structure that batches data into time-stamped blocks.

A. Blockchain security methods

B. Data stored on the blockchain

C. Proponents of permissioned

D. An IMF staff discussion

7. Because all early blockchains were permissionless, controversy has arisen over ________.

A. computer crackers

B. the blockchain definition

C. their digital assets

D. blockchain network

五、请将下列短文译成中文

1. Blockchain is a term used in information technology. In essence, it is a shared database, in which the data or information is stored, with such characteristics as "non-forgery" "trace" "traceable" "open and transparent" and "collective maintenance". Based on these characteristics, blockchain technology has laid a solid foundation of "trust", created a reliable "cooperation" mechanism, and has a broad application prospect.

2. Once recorded, the data in any given block cannot be altered retroactively without alteration of all subsequent blocks, which requires consensus of the network majority. Although blockchain records are not unalterable, blockchains may be considered secure by design. Decentralized consensus has therefore been claimed with a blockchain.

3. The bitcoin design has inspired other applications, and blockchains which are readable by the public are widely used by cryptocurrencies. Blockchain is considered a type of payment. Private blockchains have been proposed for business use. Sources such as *Computerworld* called the marketing of such blockchains without a proper security model "snake oil".

4. Nakamoto improved the design in an important way using a Hashcash-like

method to add blocks to the chain without requiring them to be signed by a trusted party. The design was implemented the following year by Nakamoto as a core component of the cryptocurrency bitcoin, where it serves as the public ledger for all transactions on the network.

5. A public blockchain has absolutely no access restrictions. Anyone with an Internet connection can send transactions to it as well as become a validator (i.e., participate in the execution of a consensus protocol). Some of the largest, most known public blockchains are the bitcoin blockchain and the Ethereum blockchain. A private blockchain is permissioned.

6. Open blockchains are more user-friendly than some traditional ownership records, which, while open to the public, still require physical access to view. Because all early blockchains were permissionless, controversy has arisen over the blockchain definition. An issue in this ongoing debate is whether a private system with verifiers tasked and authorized (permissioned) by a central authority should be considered a blockchain.

7. Blockchain technology can be integrated into multiple areas. The primary use of blockchains today is as a distributed ledger for cryptocurrencies, most notably bitcoin. Most cryptocurrencies use blockchain technology to record transactions. For example, the bitcoin network and Ethereum network are both based on blockchain. Blockchain-based smart contracts are proposed contracts that could be partially or fully executed or enforced without human interaction.

参考译文

区块链

区块链是一个信息技术领域的术语。从本质上讲,它是一个共享数据库,存储于其中的数据或信息,具有不可伪造、全程留痕、可以追溯、公开透明、集体维护等特征。基于这些特征,区块链技术奠定了坚实的"信任"基础,创造了可靠的"合作"机制,具有广阔的运用前景。

区块链,最初称作块链,是一个不断增长的称为块的记录列表,它使用密码进行链接。每个块包含前一个块的加密散列、时间戳和交易数据(通常表示为Merkle树)。区块链是被有意设计成的不可修改的数据。它是"一个开放的、分布式的账本,可以有效地、可验证地、永久地记录双方之间的交易"。作为一个分布式账本,区块链通常由一个P2P网络共同管理,它遵守节点间通信和验证新块的协议。

一旦被记录下来,任何给定块中的数据都不能在不改变所有后续块的情况下进行回溯性修改,这需要网络中多数人的一致同意。虽然区块链记录不是不可改变的,但区块链的设计可以被认为是安全的。区块链是由一个使用中本聪(Satoshi Nakamoto)名字的人在2008年发明的,是用于加密货币比特币的公共交易账簿。中本聪的身份尚不清楚。比特币区块链的发明使其成为第一种不需要可信的权威或中央服务器就能解决双消费问题的数字货币。

比特币的设计启发了其他应用,公众可读的区块链被加密货币广泛使用。区块链被认为是一种支付方式。私有区块链已经被提议用于商业用途。来自《计算机世界》的消息称,这种没有适当安全模型的区块链营销是"蛇油"。

1991年,斯图尔特·哈伯(Stuart Haber)和W.斯科特·斯托内塔(W. Scott Stornetta)描述了第一个使用加密技术保护区块链的工作。他们想要实现一个不能篡改文档时间戳的系统。在1992年,Bayer、Haber和Stornetta将Merkle树纳入设计中,将多个文档证书收集到一个块中从而提高了其效率。中本聪在一个重要方面改进了设计,使用类似Hashcash的方法将块添加到链中,而不需要由信任的一方签名。该设计于2009年由中本聪实施,作为加密货币比特币的核心组成部分,在比特币中,它是网络上所有交易的公共总账。

目前,有三种类型的区块链网络——公共区块链、私有区块链和联盟区块链。

公共区块链绝对没有访问限制。任何有互联网连接的人都可以向它发送交易,同时也可以成为一个验证器(参与协商一致的协议的执行)。一些最大、最知名的公共区块链是比特币区块链和以太坊区块链。

私有区块链要得到许可才能加入。除非受到网络管理员的邀请,否则不能加入。参与者和验证器访问受到限制。对于那些对区块链技术感兴趣但又不适应公共网络控制水平的公司来说,这种类型的区块链可以被认为是一个中间地带。通常情况下,他们寻求将区块链纳入他们的会计和记录程序,而不牺牲自主权并不用冒着将敏感数据暴露给公共互联网的风险。

一个联合区块链经常被认为是半分散管理的。它也要得到允许才能使用,但不是由单个组织来控制它,而是由许多公司在这样的网络上各自操作一个节点。联合链的管理员根据用户的意愿限制用户的阅读权限,只允许有限的可信节点执行一致协议。

区块链是一种去中心的、分布式的、公共的数字账本,它用于记录跨许多计算机的交易,因此任何涉及的记录都不能在不更改所有后续块的情况下进行回溯性更改。这允许参与者独立地、相对廉价地验证和审计交易。区块链数据库使用对等网络和分布式时间戳服务器进行自主管理。区块链的使用消除了数字资产无限再生的特性。它证实了每价值单位只转移一次,解决了长期存在的重复支出问题。区块链被描述为一种价值交换协议。区块链可以维护所有权,因为当正确地设置交换协议的细节时,它提供了一个记录来强制要约与承诺。

通过点对点网络存储数据,区块链消除了数据集中存储带来的一些风险。点对点区块链网络缺乏计算机黑客可以利用的集中弱点;同样,它也没有中心故障点。区块链安全方法包括使用公钥加密。公钥(一长串看起来随机的数字)是区块链上的一个地址。通过网络发送的价值令牌被记录为属于该地址。私钥就像密码一样,允许其所有者访问其数字资产。存储在区块链上的数据通常被认为是不可破坏的。

开放的区块链比一些传统的所有权记录更容易使用,这些记录虽然对公众开放,但仍然需要物理访问才能查看。因为所有早期的区块链都是无须许可的,所以关于区块链定义的争论就产生了。在这场正在进行的争论中，有一个问题是，一个拥有由中央授权的验证者的私有系统是否应该被认为是一个区块链。许可链或私有链的支持者认为，术语“区块链”可能适用于将数据批处理为时间戳块的任何数据结构。对于开放的、无须许可的或公共的区块链网络来说，最大的优点是不需要防范坏人，也不需要访问控制。这意味着使用区块链作为传输层可以将应用程序添加到网络中，而不需要其他人的批准或信任。

区块链技术可以结合到多个领域。如今，区块链的主要用途是作为加密货币的分布式账本，最显著的就是比特币。大多数加密货币使用区块链技术来记录交易。例如，比特币网络和以太坊网络都是基于区块链的。

基于区块链的智能合同是一种建议的合约，它可以部分或全部执行，也可以在没有人工交互的情况下强制执行。智能合同的主要目标之一是自动托管。IMF 员工讨论报告称，基于区块链技术的智能合同可能会减少道德风险，并在总体上优化合同的使用。但“目前还没有出现可行的智能合同系统。”由于缺乏广泛使用，其法律地位尚不清楚。

金融业的主要部分正在将分布式账本应用于银行业务中。银行之所以对这项技术感兴趣，是因为它有可能加快后台结算系统的速度。一些电子游戏是基于区块链技术的。第一个这样的游戏是猎手币，于 2016 年 2 月发布。另一个区块链游戏是加密猫，于 2017 年 11 月发布。这款游戏在 2017 年 12 月成为头条新闻，当时一个名为“游戏虚拟宠物”的加密猫角色以 10 万美元的价格售出。有一些行业组织致力于在供应链物流和供应链管理中采用区块链。

UNIT 16

PASSAGE

Software-Defined Networking (SDN)[1]

With the advent of cloud computing, many new networking concepts have been introduced to simplify network management and bring innovation through network programmability.[2] The emergence of the software-defined networking (SDN) paradigm is one of these adopted concepts in the cloud model so as to eliminate the network infrastructure maintenance processes and guarantee easy management.

In this fashion, SDN offers real-time performance and responds to high availability requirements. However, this new emerging paradigm has been facing many technological hurdles; some of them are inherent, while others are inherited from existing adopted technologies.

Software-defined networking (SDN) is facilitating organizations to deploy applications and enable flexible delivery, offering the capability of scaling network resources in lockstep with application and data needs, and reducing both CapEX and OpEX.[3] SDN is an innovative approach to design, implement, and manage networks that separate the network control (control plane) and the forwarding process (data plane) for a better user experience.[4] This network segmentation offers numerous benefits in terms of network flexibility and controllability.

On the one hand, it allows to combine the advantages of system virtualization and cloud computing and on the other hand, to create an implementation of a centralized intelligence that enables making a clear visibility over the network for the sake of easy network management and maintenance as well as enhanced network control and reactivity. In fact, the variety and the complexity of network elements make their maintenance very expensive and the underlying infrastructure is less reliable in case of frequent network failures, especially if no backup plans are anticipated within the infrastructure.[5]

As SDN separates the routing and forwarding decisions of networking elements (e.g., routers, switches, and access points) from the data plane, the network administration and management become uncomplicated because the control plane only deals with the information related to logical network topology, the routing of traffic, and so on. In contrast, the data plane orchestrates the network traffic in accordance with the established configuration in the control plane.

In SDN, the control operations are centralized in a controller that dictates the network policies.[6] Many controller platforms are open source such as Floodlight, OpenDaylight, and Beacon. The management of the network can be achieved on different layers (i.e., application, control, and data plane). For instance, service providers can allocate resources to customers via application layer, configure and modify network policies and logical entities on control plane, and set up physical network elements on data plane.

Unlike conventional IP networks whose functionalities are decentralized, SDN is centralized to offer connection network domains between the control and data plane in the same infrastructure.[7] Additionally, SDN allows backward compatibility with existing protocols and standards.

The core architecture of SDN is divided into three layers. The upper layer of SDN architecture is an application layer that defines rules and offers different services such as firewall, access control, quality of service, routing and proxy service.[8] This layer is responsible for abstracting the SDN network control management through the northbound API. The second layer is known as the control plane, which is an abstraction of the network topology.[9] The controller is the main component responsible for establishing flow tables and data handling policies as well as abstracting the network complexity and collecting network information through the southbound API. The lowest layer, which is known as the data plane, provides networking devices such as physical/virtual switches, routers, and access points and is responsible for all data activities including forwarding, fragmentation, and reassembly.[10]

Software-defined networking has come to light in recent years. However, the concept of this approach has been evolving since the mid-1990s. Ethane (management architecture) and OpenFlow (protocol for network flow) have given birth to a real implementation of SDN. OpenFlow is a protocol that provides a standardized way of managing traffic and describes how a controller communicates with network devices like switches and routers.[11] The devices supporting OpenFlow consist of two logical components: a flow table that defines how to process and forward packets within the network and an exposed OpenFlow application programming interface (API) that handles the exchanges between switch/router and controller.[12]

Furthermore, the demand for cloud services in its various forms is increasing drastically. Although these services are centralized in data centers, they pose important challenges for service providers. With the rapid growth of the clients' demands, the operator is required to respond accordingly by considering additional servers, network components, high quality of service, and secure architecture abiding by the standards. This generally comes first at the cost of non-negligible effort in facing new challenges appearing within the core network.

As SDN is a new networking approach, several solutions to classical network problems have been revisited using this architecture, and many problems continue to be challenging. We tried to simplify and explain some SDN issue and provided an overall view. It is important to note that the landscape of SDN-related issues changes according to the advances in SDN development. For instance, a new design modification or protocol introduced to SDN API might bring a new solution and at the same time incur new challenges or issues. Furthermore, many challenges in SDN still need further research attention such as the standardization of SDN components and adoption of new specific protocols designed to SDN in order to avoid inherited problems from the legacy networks. Research must focus more on the control plane to come up with novel solutions for controllers, which are the brains of the SDN architecture. The control plane is a point of failure of the whole network, and many security measures should be considered.

NEW WORDS AND PHRASES

advent	*n*. 到来,出现
simplify	*vt*. 简化,使简易
innovation	*n*. 创新,革新
programmability	*n*. 可编程性
emergence	*n*. 出现,浮现
define	*vt*. 定义,限定
paradigm	*n*. 范例,范式
eliminate	*vt*. 消除,排除
infrastructure	*n*. 基础设施,公共建设
guarantee	*vt*. 保证,担保
availability	*n*. 可用性
hurdle	*n*. 障碍,栏
inherent	*n*. 固有的,内在的
inherit	*v*. 继承,遗传而得
separate	*v*. 使分离,使分开
segmentation	*n*. 分割,分裂
numerous	*a*. 许多的,很多的
flexibility	*n*. 灵活性,适应性
controllability	*n*. 可控性
virtualization	*n*. 虚拟化
implementation	*n*. 实现,履行
intelligence	*n*. 智力,智能
visibility	*n*. 能见度,可见性

reactivity	*n*.反应性，活动性
conventional	*a*.符合习俗的，传统的
configuration	*n*.配置，结构
troubleshooting	*n*.故障排除
intervention	*n*.介入，干预
provision	*n*.规定，拨备　*v*.供给食物及必需品
complexity	*n*.复杂，复杂性
failure	*n*.失败，故障
especially	*ad*.特别，尤其
anticipate	*v*.预期，期望

NOTES

1. SDN，题目译为软件定义网络。

2. With the advent of cloud computing 为介词短语，意为随着云计算的出现；to simplify network management and bring innovation through network programmability 是由 and 连接的不定式短语，表示目的。本句译文：随着云计算的出现，为了简化网络管理和通过网络编程化带来创新，引入了许多新的网络概念。

3. offering the capability of scaling network resources in lockstep with application and data needs, and reducing both CapEX and OpEX，由 and 连接现在分词短语，表示伴随情况；CapEX and OpEX，资本支出和运营成本。本句译文：SDN 正在使部署应用和支持灵活交付的网络组织更加便利，提供了与应用程序和数据需求同步扩展网络资源的能力，并降低了资本支出和运营成本。

4. that separate the network control (control plane) and the forwarding process (data plane) for a better user experience，由 that 引导定语从句，对前面部分进行说明。本句译文：SDN 是设计网络、实施网络和管理网络的革新方法，它为更好的用户体验分离了网络控制(控制平面)和转发过程(数据平面)。

5. and 连接两个并列句；前面一句的名词短语做主语；in case of，假如，如果；if no backup plans are anticipated within the infrastructure, if 引导条件状语从句。本句译文：事实上，网络单元的多样性和复杂性使网络的维护非常昂贵；面对经常的网络故障，基础网络结构变得不再可靠，特别是在网络基础结构中没有补救方案时。

6. 本句为被动句；that 引导定语从句，修饰前面的 controller。本句译文：在 SDN 中，控制操作集中在一个指示网络策略的控制器中。

7. 主句是被动句；whose 引导定语从句，它是关系代词 who 的所有格。本句译文：不同于功能分散的传统 IP 网络，SDN 在同一个基础结构中进行集中管理以便在控制平面和数据平面之间提供网络域连接。

8. that 引导定语从句，修饰前面的 application layer; such as，例如。本句译文：SDN 结构的上层是应用层，用来确定规则和提供不同的服务，如防火墙、接入控制、服务质量、路由和代理服务。

9. which 引导非限定性定语从句,修饰前面部分。本句译文:称作控制平面的第二层是网络拓扑结构的抽象结果。

10. which 引导非限定性定语从句,修饰前面部分。本句译文:称作数据平面的最下一层提供联网设备如实体/虚拟交换机、路由器和接入点,并且负责数据转发、分段和重组等所有数据活动。

11. that 引导定语从句,修饰前面的 protocol;动名词短语 managing traffic 做介词 of 的宾语。本句译文:OpenFlow 是一种提供标准化的流量管理方式的协议,它描述了控制器如何与交换机和路由器等网络设备通信。

12. supporting OpenFlow 现在分词短语做后置定语,修饰前面的 devices;that 引导定语从句,修饰前面的 table;that 引导定语从句,修饰前面的 interface。本句译文:支持 OpenFlow 的设备由两个逻辑组件组成:一个是定义如何在网络中处理和转发数据包的流表;一个是公开的 OpenFlow 应用程序编程接口(API),处理交换机/路由器和控制器之间的交换。

EXERCISES

一、请将下述词组译成英文

1. 云计算的出现
2. 简化网络管理
3. SDN 架构的上层
4. 定义规则的应用层
5. 网络拓扑的抽象
6. 流表和数据处理的策略
7. 减少 CapEX 和 OpEX
8. 许多技术障碍
9. 高可用性需求
10. 确保方便管理
11. 从现有技术继承
12. 网络控制和转发过程
13. 这种网络分割
14. 网络灵活性和可控性
15. 系统虚拟化和云计算
16. 网络设备的复杂性
17. 经常的网络故障
18. 流量的路由

二、请将下述词组译成中文

1. to eliminate the network maintenance processes
2. in terms of network flexibility and controllability
3. the emergence of the software-defined networking (SDN)
4. the advantages of virtualization and cloud computing
5. for the sake of easy network management
6. the variety and the complexity of network elements
7. in case of frequent network failures
8. to separates the routing and forwarding decisions
9. the information related to logical network topology
10. the established configuration in the control plane
11. the controller that dictates the network policies
12. to allocate resources to customers via application layer
13. to set up physical network elements on data plane
14. backward compatibility with existing protocols

15. to collect network information through the southbound API

16. to provides a standardized way of managing traffic

三、选择合适的答案填空

1. The emergence of the software-defined networking (SDN) paradigm is one of these adopted concepts in the cloud model so as ________ the network infrastructure maintenance processes and guarantee easy management.

A. to eliminate　　B. eliminating

C. eliminated　　D. eliminate

2. However, this new emerging paradigm has been facing many technological hurdles; some of them are inherent, while others are ________ from existing adopted technologies.

A. inheriting　　B. to inherit

C. inherit　　D. inherited

3. Software-defined networking has come to light in recent years. However, the concept of this approach has been ________ since the mid-1990s.

A. evolve　　B. evolving

C. evolved　　D. to evolve

4. On the one hand, it allows to combine the advantages of system virtualization and cloud computing and on the other hand, ________ an implementation of a centralized intelligence that enables making a clear visibility over the network for the sake of easy network management and maintenance as well as enhanced network control and reactivity.

A. create　　B. created

C. to create　　D. creating

5. The core architecture of SDN is divided into three layers. The upper layer of SDN architecture is an application layer that defines rules and offers different services such as firewall, access control, quality of service, routing and proxy service. This layer is responsible for ________ the SDN network control management through the northbound API.

A. abstract　　B. to abstract

C. abstracted　　D. abstracting

6. The controller is the main component responsible for establishing flow tables and data handling policies as well as ________ the network complexity and collecting network information through the southbound API.

A. abstract　　B. abstracting

C. to abstract　　D. abstracted

四、根据课文内容选择答案

1. The emergence of the software-defined networking (SDN) paradigm is one of

these adopted concepts in the cloud model so as to eliminate the network infrastructure maintenance processes and ________.

A. guarantee easy management

B. offers real-time performance

C. better user experience

D. offers numerous benefits

2. However, this new emerging paradigm has been facing many technological hurdles; some of them are inherent, while others are ________.

A. introduced to simplify network management

B. adopted concepts in the cloud model

C. enhanced network control and reactivity

D. inherited from existing adopted technologies

3. This network segmentation offers numerous benefits in terms of ________.

A. system virtualization and cloud computing

B. network flexibility and controllability

C. application and data needs

D. existing adopted technologies

4. In fact, the variety and the complexity of network elements make their maintenance very expensive and the underlying infrastructure is less reliable in case of frequent network failures, especially if ________ no within the infrastructure.

A. SDN is an innovative approach

B. network segmentation offers numerous benefits

C. backup plans are anticipated

D. networking concepts have been introduced

5. In contrast, the data plane orchestrates the network traffic in accordance with the established configuration ________.

A. within the infrastructure

B. in the control plane

C. in terms of network flexibility

D. in the cloud model

6. The controller is the main component responsible for ________ and data handling policies as well as abstracting the network complexity and collecting network information through the southbound API.

A. establishing flow tables

B. network control management

C. routing and forwarding decisions

D. facilitating organizations to deploy applications

7. The lowest layer, which is known as the data plane, provides networking

devices such as physical/virtual switches, routers, and access points and is ________ including forwarding, fragmentation, and reassembly.

A. abstraction of the network topology

B. consist of two logical components

C. standardized way of managing traffic

D. responsible for all data activities

五、请将下列短文译成中文

1. With the advent of cloud computing, many new networking concepts have been introduced to simplify network management and bring innovation through network programmability. The emergence of the software-defined networking (SDN) paradigm is one of these adopted concepts in the cloud model so as to eliminate the network infrastructure maintenance processes and guarantee easy management.

2. In this fashion, SDN offers real-time performance and responds to high availability requirements. However, this new emerging paradigm has been facing many technological hurdles; some of them are inherent, while others are inherited from existing adopted technologies.

3. On the one hand, it allows to combine the advantages of system virtualization and cloud computing and on the other hand, to create an implementation of a centralized intelligence that enables making a clear visibility over the network for the sake of easy network management and maintenance as well as enhanced network control and reactivity.

4. In SDN, the control operations are centralized in a controller that dictates the network policies. Many controller platforms are open source such as Floodlight, OpenDaylight, and Beacon. The management of the network can be achieved on different layers (i.e., application, control, and data plane).

5. The second layer is known as the control plane, which is an abstraction of the network topology. The controller is the main component responsible for establishing flow tables and data handling policies as well as abstracting the network complexity and collecting network information through the southbound API.

6. The lowest layer, which is known as the data plane, provides networking devices such as physical/virtual switches, routers, and access points and is responsible for all data activities including forwarding, fragmentation, and reassembly. Software-defined networking has come to light in recent years.

7. Furthermore, the demand for cloud services in its various forms is increasing drastically. Although these services are centralized in data centers, they pose important challenges for service providers.

8. As SDN is a new networking approach, several solutions to classical network problems have been revisited using this architecture, and many problems continue to be

challenging. We tried to simplify and explain some SDN issue and provided an overall view.

参考译文

软件定义网络

随着云计算的出现,为了简化网络管理和通过网络编程化带来创新,引入了许多新的网络概念。软件定义网络SDN实例的出现就是云模式中的概念之一,其目的是消除网络结构的维护过程和保证更容易的网络管理。

在SDN方式中,SDN提供实时性能并对高可用性要求进行响应。然而,这种SDN新应用要面临许多技术障碍,一些是SDN本身固有的,而其他障碍源于现有继承的技术。

SDN正在使部署应用和支持灵活交付的网络组织更加便利,提供了与应用程序和数据需求同步扩展网络资源的能力,并降低了资本支出和运营成本。SDN是设计网络、实施网络和管理网络的革新方法,它为更好的用户体验分离了网络控制(控制平面)和转发过程(数据平面)。这种网络分割在网络灵活性和可控性方面有许多益处。

一方面SDN可以结合系统虚拟化和云计算的优势,另一方面为了方便网络管理和维护以及提高网络控制和反应性可以实现清晰的可视化智能化中心。事实上,网络单元的多样性和复杂性使网络的维护非常昂贵,面对经常的网络故障,基础网络结构变得不再可靠,特别是在网络基础结构中没有补救方案时。

由于SDN将网络单元(如路由器、交换机和接入点)的路由和转发决策从数据平面分离出来,网络管理变得简单,因为控制平面只处理与逻辑网络拓扑、流量路由等相关的信息。相反,数据平面根据控制平面中已建立的配置来编排网络流量。

在SDN中,控制操作集中在一个指示网络策略的控制器中。许多控制器平台都是开源的,比如Floodlight、OpenDaylight和Beacon。网络的管理可以在不同的层(即应用程序、控制和数据平面)实现。例如,业务提供者可以通过应用层向客户分配资源,在控制平面上配置和修改网络策略和逻辑实体,在数据平面上设置物理网络单元。

不同于功能分散的传统IP网络,SDN在同一个基础结构中进行集中管理以便在控制平面和数据平面之间提供网络域连接。此外,SDN允许与现有的协议和标准后向兼容。

SDN的核心结构分为三层。SDN结构的上层是应用层,用来确定规则和提供不同的业务,如防火墙、接入控制、服务质量、路由和代理服务。这一层通过北向API负责对SDN网络控制管理进行抽象。称作控制平面的第二层是网络拓扑结构的抽象结果。其中的控制器是主要部件,用来建立流表和数据处理策略以及对网络复杂性进行抽象,还可以通过南向API收集网络信息。称作数据平面的最下一层提供联网设备如实体/虚拟交换机、路由器和接入点,并且负责数据转发、分段和重组等所有数据活动。

软件定义网络近年来逐渐为人所知。然而,自20世纪90年代中期以来,这种方法的概念一直在演变。Ethane(管理体系结构)和OpenFlow(网络流协议)催生了SDN的真正实现。OpenFlow是一种提供标准化的流量管理方式的协议,它描述了控制器如何与交换机和路由器等网络设备通信。支持 OpenFlow 的设备由两个逻辑组件组成:一个是定义如何在网络中处理和转发数据包的流表;一个是公开的 OpenFlow 应用程序编程接口(API),用来处理交换机/路由器和控制器之间的交换。

此外,对各种形式的云服务的需求也在急剧增加。尽管这些服务集中在数据中心,但它们对服务提供者构成了重要的挑战。随着客户需求的快速增长,运营商需要考虑额外的服务器、网络组件、高质量的服务以及符合标准的安全架构,并对此做出相应的响应。通常这首先要付出一定的努力为代价,以面对核心网络中出现的新挑战。

由于 SDN 是一种新的网络方法,使用这种体系结构对传统网络问题的几种解决方案进行了重新探讨,许多问题仍然具有挑战性。我们试图简化和解释一些 SDN 问题,并提供了一个总体的观点。值得注意的是,SDN 相关问题的前景会随着 SDN 的发展而变化。例如,引入 SDN API 的新修改的设计或协议可能带来新的解决方案,同时带来新的挑战或问题。此外,SDN 网络中的许多挑战还需要进一步的研究,如 SDN 组件的标准化和采用新的专门为 SDN 设计的协议,以避免继承网络的遗留问题。研究必须更多地集中在控制平面,为控制器提出新的解决方案,这是 SDN 架构的核心。控制平面是整个网络的一个故障点,需要考虑多种安全措施。

UNIT 17

PASSAGE

Introduction to Optical Fiber Communication[1]

One of the most important technological developments during the 1980s has been the emergence of optical fiber communication as a major international industry. One indication of the extent of this development is the total length of installed fiber, which was estimated to be 3.2 million kilometers in the U.S. alone by the end of 1987.[2] Over 90% of this fiber was placed in service during the time period of 1982 – 1987. Long-haul trunk installations have been dominated, accounting for about 95% of the fiber in the U. S.

Although telecommunication is the rationale for most of the current interest in fiber optics, this was not the case during the early days of the technology. The researchers who produced the first clad glass optical fibers in the early 1950s were not thinking of using them for communications; they wanted to make imaging bundles for endoscopy. Fiber optics was already a well-established commercial technology when the famous paper by Kao and Hockham, suggesting the use of low-loss optical fibers for communication, appeared in 1966.[3]

The first low-loss (20 dB/km) silica fiber was described in a publication which appeared in October of 1970. The date of this publication is sometimes cited as the beginning of the era of fiber communication. Although this development did receive[4] considerable attention in the research community at the time, it was far from inevitable that a major industry would evolve. The 20 dB/km loss figure was still too high for long-haul telecommunication systems. The fibers were fragile, and a way to protect them would have to be found. There were no suitable light sources. Researchers did not know whether field termination and splicing of optical cables would ever be practical. Finally, there were serious doubts as to whether these components could ever be produced[5] economically enough for the technology to play a major role in the marketplace.

Although the technological barriers appeared formidable, the economic potential was very significant. As a consequence, research and development activity expanded rapidly, and a number of important issues were resolved during the early 1970s.

During the middle and late 1970s, the rate of progress toward marketable products accelerated as the emphasis shifted from research to engineering.[6] Fibers with losses approaching the Rayleigh limit[7] of 2 dB/km at a wavelength of 0.8 μm, 0.3 dB/km at 1.3 μm, and 0. 15 dB/km at 1. 55 μm, were produced in the laboratory. Microbend loss problems were overcome through the use of improved fiber coatings and cabling techniques. Rugged cables and multifibre connectors were produced for field installation. Room temperature threshold currents for commercial gallium aluminum arsenide lasers operating in the 0.8 to 0.85 μm spectral region were reduced to the 20 to 30 mA range, and projected lifetimes in the 100 000 to 1 000 000 hour range were claimed for both lasers and LEDs. Light sources and improved photodetectors which operated near 1.3 μm were developed to take advantage of the low fibre loss and dispersion in this "longer wavelength region". Several major field trials were undertaken during this period, including AT&T's Atlanta experiment and Chicago installation, and Japan's subscriber access project.

Improvements in component performance, cost, and reliability by 1980[8] led to major commitments on the part of telephone companies. Fiber soon became the preferred transmission medium for long-haul trunks. Some early installations used 0.8 μm light sources and graded-index multimode fiber, but by 1983, designers of intercity links were thinking in terms of 1.3 μm, single-mode systems. The single-mode fibre, used in conjunction with a 1.3 μm laser, provides a bandwidth advantage which translates into increased repeater spacings for high data rate systems.

Data rates for installed fiber optic systems have recently moved into gigabit per second range. Such systems use the spectrally pure distributed-feedback lasers to minimize fibre dispersion effects. Fibres designed for low dispersion at 1.55 μm wavelength, which corresponds to minimum fiber loss, are now commonly used in long distance transmission. The use of wavelength multiplexing to further increase the fibre information capacity is becoming more widespread.

The potential of fibre optics in other areas is only beginning to be realized. Fibre optic networks for computer systems and offices are becoming more prominent. In the telephone system, the use of fibre optics for interconnecting central offices within a metropolitan area and for lower levels in the switching hierarchy is still increasing rapidly. Fibre links to the home have been used in demonstration projects. Many observers believe that national telephone systems will eventually be upgraded to handle video bandwidths by using fibre optics. These wideband subscriber loop systems would provide access to services such as picturephone, video entertainment. Widespread installation of these broadband services will become economically feasible.

NEW WORDS AND PHRASES

fiber = fibre　　*n*. 光纤，纤维
emerge　　*vi*. 出现，形成，浮现
emergence　　*n*. 浮现，出现
haul　　*vt*. 用力拖；*n*. 拖，拉，拖运的距离

long-haul	长运距的，长途的
trunk	*n*. 干线，中继线，中继线路，局内线
dominate	*vt*. 支配，统治；处于支配地位
account	*n*. 账目，报道；说明
account for	解释，(数量等)占
rationale	*n*. (某事物的)基本理由，理论基础
clad	*a*. 穿衣的，(金属)包层；*n*. 覆盖层，(用壳)包盖
bundle	*n*. 捆，束
endoscopy	*n*. 内窥镜
silica	*n*. 石英，二氧化硅
publication	*n*. 发表，出版物
cite	*vt*. 引用，举(例)
community	*n*. 团体，社会
inevitable	*a*. 不可避免的，必然的，合情合理的
evolve	*vt*. 使发展，推论；进展，发展，进化
figure	*n*. 外形，图形，人物，数字
fragile	*a*. 脆的，易碎的
marketplace	*n*. 市场
barrier	*n*. 隔板，障碍，界限
formidable	*a*. 可怕的，难以应付的，庞大的
consequence	*n*. 结果，结论
accelerate	*vt*. (*vi*.) 加速
Rayleigh	*n*. 瑞利(人名)
rugged	*a*. 强壮的，艰苦的
threshold	*n*. 门槛，阀，门限
gallium aluminum arsenide (GaAlAs) lasers	镓铝砷激光器
photodetector	*n*.光电探测器，光电检测器
project	*vt*.设计，抛出；*n*. 项目，方案
reliability	*n*. 可靠性
trial	*n*. (好坏、性能等的)试验
AT&T (American Telephone and Telegraph Company)	美国电话电报公司
Atlanta	*n*. 亚特兰大(美国城市)
Chicago	*n*. 芝加哥(美国城市)

commitment	*n*. 作为，赞助，委托
intercity	*a*. 城市之间的，市际的
gigabit	*n*. 吉比特，千兆比特
spectrally-pure	*a*. 光谱纯的
distributed-feedback(DFB) laser	分布反馈激光器
potential	*n*. 潜力，可能
prominent	*a*. 突出的，杰出的，重要的
metropolitan	*a*. 首都的，主要城市的
hierarchy	*n*. 分类等级
upgrade	*vt*. 使升级
video	*a*. 使视频的，电视的
wideband	*n*. 宽频带，宽波段
entertainment	*n*. 招待，娱乐
broadband	*n*. 宽频带，宽波段
feasible	*a*. 可实行的，行得通的，可能的

NOTES

1. 本篇课文涉及光纤通信领域，题目可译为：光纤通信介绍。

2. by the end of 1987，到 1987 年底时。其中 by 为介词，意思为“不迟于”“到……时(为止)”。

3. when the famous paper by Kao and Hockham, suggesting the use of low-loss optical fibres for communication, appeared in 1966. 这是一个时间状语从句，从句中主语和谓语割裂，主语为 the famous paper，谓语为 appeared。该句可译成：1966 年，当高锟和霍克姆的著名论文提出在通信中使用低损耗光纤时，光纤已经是一种成熟的商业技术。

4. although this development did receive … 这是一个让步状语从句，谓语动词 receive 前加 did 是为了强调。

5. there were serious doubts as to whether these components could ever be produced … 这一句中的 as to 是个短语，与 as regards 相同，意为“至于……”，它们经常与 doubt、agree 等词连用，后跟 whether、what、how 等词引出的从句。句中的 ever，意思为“究竟，到底，非常”，用来加强语气。

6. as the emphasis shifted from research to engineering，这是原因状语从句，句中的 shift from … to … 意为“由……转移到……”。

7. Rayleigh limit 瑞利极限(在光纤中，瑞利散射损耗是固有的，不可消除的，因而瑞利散射是光纤散射损耗的最低极限)。

8. by 1980，到 1980 年时，到 20 世纪 80 年代时。

EXERCISES

一、请将下述词组译成英文

1. 光纤通信　2. 光源　3. 波长　4. 激光器
5. 色散　6. 传输介质　7. 多模光纤　8. 长途干线
9. 单模光纤　10. 带宽　11. 宽带用户　12. 纤维光学
13. 商用技术　14. 门限电流　15. 光检测器　16. 波分复用
17. 纤维光网络　18. 视频带宽

二、请将下述词组译成中文

1. long distance transmission
2. repeater spacing
3. commercial technology
4. optical fibre communications
5. the total length of installed fibre
6. long-haul telecommunication system
7. the low-loss silica fibre
8. fibers with losses approaching the Rayleigh limit
9. room temperature threshold currents
10. the longer wavelength region
11. subscriber access project
12. improvements in component performance and reliability
13. data rates for installed fibre optic system
14. gigabit per second range
15. wavelength multiplexing
16. wideband subscriber loop system
17. multifibre connectors
18. projected lifetime
19. light source
20. single-mode fibre
21. distributed-feedback laser
22. information capacity
23. switching hierarchy
24. broadband services

三、选择合适的答案填空

1. One indication of the extent of this development is the total length of ________ fiber, which was estimated ________ 3.2 million kilometers in the U.S.

A. install, be　　B. installing, being
C. to install, was　　D. installed, to be

2. The researchers ________ produced the first clad glass optical fibers were not thinking of ________ them for communications.

A. which, use　　B. that, used

C. who, using　　D. one, to use

3. The single-mode fiber, ________ in conjunction with a 1.3 μm laser, provides a bandwidth advantages which translates into ________ repeater spacings for high-data rate systems.

A. used, increased　　B. using, increase

C. use, to increase　　D. to use, increasing

4. In the telephone system, the use of fiber optics for ________ central offices within a metropolitan area and for lower levels in the switching hierarchy is still ________ rapidly.

A. interconnect, increase　　B. interconnecting, increasing

C. to interconnect, to increase　　D. interconnected, increased

5. The fibres were fragile, and a way ________ would have ________ .

A. protect, found　　B. protection, find

C. to protect, to be found　　D. protecting, finding

6. Many observers believe that national telephone system will eventually be upgraded ________ video bandwidths by ________ fibre optics.

A. handle, use　　B. handled, used

C. handling, usage　　D. to handle, using

四、根据课文内容选择正确答案

1. The first low-loss silica fiber was described in a publication which appeared in October of 1970. The date of this publication is sometimes cited as ________.

A. the beginning of the era of fiber communication

B. the start of the data communicating

C. the beginning of the new era of telecommunication

D. the start of the era of information

2. Data rates for installed fiber optic systems have recently moved into ________.

A. the gigabit per second range

B. the million bits per second range

C. the 8 digits per second range

D. the rate of 64 kHz

3. The use of wavelength multiplexing ________ is becoming more widespread.

A. to increase the fiber information capacity

B. to decrease the fiber dispersion

C. to interconnect the switching centers

D. to handle video bandwidth

4. Fiber links to the home have been used ________ .

A. everywhere

B. in demonstration projects

C. in the gigabit per second range

D. by single-mode systems

5. The wideband subscriber loop systems would provide access to services ________.

A. such as telephones

B. such as the low-speed data communications

C. such as picturephone

D. such as switching

6. One of the most important technological developments during the 1980s has been the emergence of ________ as a major international industry.

A. radio communication

B. switching technology

C. voice communication

D. optical fibre communication

五、请将下述短文译成中文

1. Although telecommunication is the rationale for most of the current interest in fibre optics, this was not the case during the early days of the technology. The researchers who produced the first clad glass optical fibres in the early 1950s were not thinking of using them for communications; they wanted to make imaging bundles for endoscopy. Fibre optics was already a well established commercial technology when the famous paper by Kao and Hockham, suggesting the use of low-loss optical fibres for communication, appeared in 1966.

2. The first low-loss silica fibre was described in a publication which appeared in October of 1970. The date of this publication is sometimes cited as the beginning of the era of fibre communication. Although this development did receive considerable attention in the research community at the time, it was far from inevitable that a major industry would evolve.

3. The 20 dB/km loss figure was still too high for long-haul telecommunication systems. The fiber was fragile, and a way to protect them would have to be found. There were no suitable light sources. Researchers did not know whether field termination and splicing of optical cables would ever be practical. Finally, there were serious doubts as to whether these components could ever be produced economically enough for the technology to play a major role in the marketplace.

4. During the middle and late 1970s, the rate of progress toward marketable products accelerated as the emphasis shifted from research to engineering. Fibers with losses approaching the Rayleigh limit were produced in the laboratory. Microbend loss

problems were overcome through the use of improved fiber coatings and cabling techniques. Light sources and improved photodetectors which operated near 1.3 μm were developed to take advantage of the low fiber loss and dispersion in this “longer wavelength region”.

5. Data rates for installed fiber optic systems have recently moved into the gigabit per second range. Such systems use the spectrally pure distributed-feedback lasers to minimize fiber dispersion effects. Fibers designed for low dispersion at 1.55 μm wavelength, which corresponds to minimum fiber loss, are now commonly used in long distance transmission. The use of wavelength multiplexing to further increase the fiber information capacity is becoming more widespread.

6. The potential of fiber optics in other areas is only beginning to be realized. Fiber optic networks for computer systems and offices are becoming more prominent. In the telephone system, the use of fiber optics for low levels in the switching hierarchy is still increasing rapidly. Fiber links to the home have been used in demonstration projects. Many observers believe that national telephone systems will eventually be upgraded to handle video bandwidths by using fiber optics.

7. Three components are involved in a basic optical fiber system: the light source, the photodetector, and the optical transmission line. The optical light source generates the optical energy which serves as the information carrier, similar to a radio wave source supplying electromagnetic energy at radiowave wavelengths as the information carrier. The optical photodetector detects the optical energy and converts it into an electrical form. The optical fiber transmission line is the equivalent of copper wires and functions as the conductor of optical energy.

参考译文

光纤通信介绍

20世纪80年代一项最重要的技术发展是光纤通信成为一个主要的国际性产业。用光纤敷设总长度可以表明其发展程序。据估计，到1987年底仅美国的光纤敷设总长将达320万千米，其中90%以上是在1982—1987年间敷设并开通的，而长途干线占主导地位，数量约为光纤总长的95%。

虽然现在人们对纤维光学的兴趣主要在于通信，但早期发展纤维光学的目的并不在此。20世纪50年代初研究人员制造出第一根包层玻璃光纤时，并不想将其用于通信，而是想用它们传送内窥镜需要的成像光束。1966年Kao和Hockham发表了那篇著名的论文，建议将低损耗光纤用于通信，此时纤维光学已发展为一项很实用的技术了。

1970年10月，第一根低损耗(20 dB/km)石英光纤问世了。有时我们将这一日期作为光纤通信时代的开端。虽然这一成果当时在研究领域确实引起了极大的关注，但这种光纤距离通信所要求的条件还相差甚远：每千米20 dB的损耗对于长途通信系统仍然是太大了；光纤易断裂，必须寻找保护方法；没有合适的光源。研究人员不知道光缆的终端和接头是否会发展到实用阶段,对于生产这些器件在经济上是否可行,从而使之在市场上占有重要地位,他们更是存有严重的疑虑。

虽然技术障碍好像不可逾越,但经济潜力却非常明显。正因为如此,20世纪70年代早期的研究和开发工作开展迅速,一些重要问题得以解决。

20世纪70年代中后期,由于发展重点由研究领域转入工程实用,因而加速了产品推向市场的速度。在实验室研制的光纤衰减值接近瑞利极限值:0.8 μm波长处为2 dB/km,1.3 μm波长处为0.3 dB/μm和1.55 μm波长处为0.15 dB/km。通过改进光纤外涂层的方法和成缆技术,克服了微弯损耗。加强型电缆和多纤连接器被生产并用于室外作业。工作在0.8~0.85 μm波长区的商用镓铝砷激光器的室温阈值电流减少到20~30 mA范围。据称,激光器和发光管的设计寿命达10万~100万小时。开发了工作于1.3 μm波长附近的光源和改进的光电检测器,从而可以利用光纤在长波长区的低损耗和低色散特性。这一时期进行的室外实验较重要的有AT&T于1976年在亚特兰大的实验、1977年在芝加哥的实验和1977年日本的用户接入项目。

到了20世纪80年代,光纤器件在性能、价格和可靠性方面的改善使众多电话公司受益匪浅。光纤很快成为长途干线的首选传输媒质。一些早期敷设的光纤线路采用0.8 μm的光源和渐变折射率多模光纤,但到1983年,城市间线路的设计者们就考虑使用1.3 μm的单模光纤系统了。单模光纤与1.3 μm激光器连接,可以提供宽带特性,增加了高速率系统的中继距离。

最近敷设的光纤系统的数据速率已移至每秒吉比特范围。这种系统采用光谱纯的分布反馈激光器,将光纤色散效应减至最小。在1.55 μm波长上设计的低色散光纤,相应地具有低损耗特性,目前广泛用于长途通信。为进一步增加光纤的信息容量,逐渐广泛采用波分复用方法。

人们对于光纤在其他领域的潜力刚刚开始认识。用于计算机系统和办公室的光纤网络逐渐变得更加重要。在电话系统中,光纤在主要城市地区中心交换局间互联和低级交换中的使用继续迅速增加。入户光缆已经有了示范工程。许多观察家相信,全国电话系统最终将升级到使用光纤传输视频宽带信号。这些宽带用户环路系统将为可视电话、视频娱乐节目等业务提供通路。宽带业务广泛使用光纤将会变得经济可行。

第四部分
练习参考答案

UNIT 1
Optical Transmission Network OTN

一、词组翻译参考答案

1. the data services developed rapidly
2. IPTV and video development
3. to provide high bandwidth
4. a transport network scheduled quickly
5. the convenient network maintenance and management
6. to meet the needs of the service
7. to use the technology of SDH and WDM
8. the electrical-layer service
9. similar to the SDH DXC equipments
10. to provide powerful OAM capabilities
11. the WDM technology based on optical layer
12. the advantages of traditional SDH
13. the broadband service of large particle

二、词组翻译参考答案

1. 灵活的调度、管理和保护机制
2. 多波长信道传输特性
3. 实现大颗粒传输
4. 在电层的异步映射和复用
5. 改善适应性和传输效率
6. 光层和电层的完整系统结构
7. 基于交叉功能 ODU*k* 的 OTN 设备
8. 引入基于 ASON 的智能控制平面
9. 改善网络配置的灵活性和生存性
10. 接口适配器和线路接口处理
11. 主流 WDM 系统制造商
12. 提供各种外部接口和 OUT*k* 接口
13. OTN 设备光电混合交叉
14. 不同 OTN 应用方法
15. 为了更有效地使用 IP 网络
16. OTN 大容量设备的引入

17. 大容量 OTN 设备的使用

18. 各种保护和恢复方法

三、选择题答案

1. D　2. A　3. C　4. B　5. C　6. A

四、选择题答案

1. B　2. D　3. A　4. C　5. B　6. D

五、参考译文

1. 近年来,通信网络所承载的业务发生了巨大的变化。数据业务发展迅速,尤其是宽带业务,如 IPTV 和视频开发。

2. 更重要的是,它需要传输网络可以快速灵活地调度、完善和方便的网络维护和管理(OAM 功能),以满足服务的需要。目前主要的传输网络是使用 SDH 和 WDM 技术,但它们都有一些局限性。

3. SDH 技术强调电层服务,具有灵活的调度、管理和保护机制以及完善的 OAM 功能。传统 SDH 传输网业务调度颗粒小,传送容量有限,不能有效地传送宽带大颗粒业务。

4. 但是,目前的 WDM 网络主要应用于点对点模式,缺乏有效的网络维护和管理工具。光调度系统(如 ROADM)可以实现类似于 SDH 的调度和保护能力,但由于物理层和波长的限制,很难应用于广泛的网络。

5. OTN 技术兼有传统 SDH 和 WDM 的优势。OTN 在光层采用 WDM 技术,可以实现大颗粒的传送。OTN 在电层使用异步的映射和复用,支持 ODUk(k=1,2,3,4,flex)的交叉连接颗粒。

6. OTN 技术包括完整的光层和电层系统结构,每层都有相应的网络管理、监控和生存机制。OTN 技术可提供强大的 OAM 功能,并支持多达六个的串联连接监控(TCM)功能,提供高级性能和故障监控。

7. OTN 设备应具有客户接口、接口适配器以及线路接口处理功能。有几种形式的 OTN 设备,如 OTN 终端复用器、OTN 电交叉设备和 OTN 光电混合交叉设备。

8. 随着远程 IP 网络的发展,IP 流量激增,远程骨干核心节点面临着越来越多的业务量,并且为了更有效地利用 IP 网络资源,提高中继电路的利用率或网络运营质量,有必要在远程骨干网中使用大容量 OTN 交叉设备。

UNIT 2
Synchronous Digital Hierarchy

一、词组翻译参考答案

1. synchronous digital hierarchy

2. international standard

3. signal format
4. network node interface
5. tributary signals
6. digital cross-connection
7. network management
8. network maintenance
9. network operators
10. transmission rate
11. tributary mapping
12. flexibility
13. subscriber services
14. overlay levels
15. manufacturer
16. synchronous transmission frame
17. line terminal multiplexer
18. add-drop multiplexer
19. regenerator
20. sensitivity
21. virtual container
22. framing byte
23. section overhead
24. end-to-end transmission
25. error monitoring
26. signal processing nodes
27. payload
28. pointer

二、词组翻译参考答案

1. 同步传输系统
2. 覆盖 NNI 的标准
3. 国际标准接口
4. 直接同步复用
5. 灵活的通信联网
6. 点对点的传输技术
7. 先进的网络管理
8. 不同厂家提供的设备
9. SDH 提供的灵活性
10. 同步复用设备的运营者
11. 电信联网

12. 支路信号
13. 维护能力
14. 统一的电信网络基础结构
15. 组件
16. 终端复用器
17. 贯通方式
18. 同步数字交叉连接
19. 可变带宽
20. 各个支路信号
21. 传输系统
22. 光载体
23. 二维图形
24. 传送次序
25. 成帧字节
26. 虚容器
27. 段开销
28. 误码检测

三、选择题答案

1. C 2. A 3. D 4. A 5. D 6. D 7. B 8. C

四、选择题答案

1. B 2. D 3. C 4. D 5. C 6. D 7. C 8. B

五、参考译文

1. 1988年11月,第一批SDH标准通过。这些标准定义了网络节点接口(NNI)的传输速度、信号格式、复用结构和支路映射。网络节点接口,即同步数字系列的国际接口。正是由于SDH设备在这些方面的标准化,才提供了网络运营者所期望的灵活性,从而能低价高效地应付带宽方面的增长并为后十年中将出现的新的用户业务做好准备。

2. SDH标准的基础是直接同步复用。这一原理是低价高效和灵活组网的关键。从本质上讲,这意味着各支路信号可以被直接复用到更高速率的SDH信号之中,而用不着中间级的复用阶段。诸SDH网络单元可以直接相互连接,因而,显然比现在的网络更省钱、更省设备。

3. SDH能够传输当今电信网中所有常见的支路信号。这意味着我们可用SDH作为现有信号类型的总包层。此外,对于网络运营者将来欲支持的各种新型的用户业务信号,SDH亦具有直接处理的灵活性。

4. PDH的局限主要表现如下:

- 1.5 Mbit/s和2 Mbit/s的PDH系列的不兼容性不利于国际通信的发展。
- 异步复用在各级码速调整、上下话路方面的不灵活限制了其向更高阶的发展和对巨大光纤传输容量的更有效的利用。
- 在PDH的帧结构中没有开销比特,造成OAM功能不好。

- PDH 的光接口，包括线路码型没有标准化，因而由不同厂家制造的设备不能互通，线路系统缺乏横向兼容能力。

5. 线路终端复用单元可接受一批支路信号并将它们复用至适当的 SDH 速率的光载体上，例如 STM-16。输入的支路信号可以是现有的 PDH 信号，也可以是较低速率的 SDH 信号。线路终端复用单元(LTM)形成了由 PDH 网络到 SDH 的主入口。

6. 分插复用单元是一种特殊类型的终端复用单元，它是以"贯通"模式运行的。在分插复用单元(ADM)中，可以从"贯通"信号中上、下话路。ADM 的功能是 SDH 的主要优势之一，因为在 PDH 网络中，完成类似功能需要一套硬布线的背对背的终端设备。

7. 同步数字交叉连接单元将成为新的同步数字系列的基石。它们能对传输信道起到半永久交换的作用，并可在从 64 kbit/s 直至 STM-1 速率的任何级别上进行交换。一般地，这种单元具有 STM-1 或 STM-4 的接口。DXC 可以在软件控制下迅速地重构电路，以提供数字租用线路或变带宽的其他业务。

8. SDH 的主要优点可表述如下：

- 网络单元价格较低：由于有公共标准，许多厂商的设备将能兼容。在市场竞争激烈的情况下，其价格将极为诱人。
- 网络管理更佳：在网管更好的情况下，运营者将能更高效地利用网络和提供更好的服务。CCITT 正在研究 TMN 的概念。一些定义管理系统接口的 TMN 标准已经存在。
- 电路的提供速度更快：如果能用软件定义新电路以便利用现有的空闲带宽，则电路的提供速度将快得多。所需要的新的连线仅为从用户的家里到最近的网络接入节点。

9. SDH 的主要优点可表述如下：

- 更好的网络利用率：在完全控制了路由选择的情况下，用户电路可以被准备和集中以便最佳利用网络资源。典型地讲，所有的话音传送电路都可与数据电路分开并以最小的时延选择路由。而数据电路可按照其类型在所需的交叉连接级别上被集中到特定的网络 DXC 上。
- 更好的网络生存性：由于实时地重选路由成为可能，网络运行支撑系统只要简单地重排电路路由，就能照顾到电路故障。内设的保护和报警系统将自动地注意到简单的传输故障。
- 更简单的交接：如果所有的网络采用遵循同样标准的设备，在网络节点接口上的电路交接将准确无误。
- 支持未来业务：展望未来，SDH 的设计将适合未来的业务，例如高清电视、广域网的骨干网、宽带 ISDN 和新的宽带点播业务。因为 SDH 的运营者将完全控制带宽分配，任何新业务都能很简单地提供。

10. 为清晰起见，在 STM-1 中的一帧可用一个二维的图形表示。该二维结构由9 行、270 列个方框组成。每个方框代表同步信号中的一个 8 比特字节。在二维图的左上角有 6 个成帧字节，这些成帧字节起着标志作用，它使帧中的任何字节极易被确定位置。

11. 一个同步传输帧由帧结构中的两个不同的、可直接接入的部分——虚容器部分和段开销部分组成。各支路信号被安排在虚容器中通过 SDH 网络进行端到端的传输。段开销提供了支持和维护同步网络节点间 VC 的传输所需要的开销。

12. 虚容器用于在同步网中传送支路信号。在多数情况下,这个信号是在同步网的入口处被组装起来,并在出口处被分拆。在同步网中,虚容器在其通过网络的路由里,无损伤地在传输系统间通行。但是,段开销仅仅属于一个单传输系统并且支持 VC 通过该传输系统传送。它不随 VC 在传输系统之间转移。

13. 为了照顾在同步网中小的定时偏差,并且简化信号的复用和交叉连接,VC-4 被允许在 STM-1 帧的净负荷区中浮动。这意味着 VC-4 可开始于 STM-1 净负荷区中的任何地方,而且一般不会整个地装在一帧之中,倒很有可能在某一帧中开始并在下一帧中结束。

UNIT 3
The Principle of PCM

一、词组翻译参考答案

1. sampling, quantizing and coding
2. speech channel
3. amplitude value
4. sampling frequency
5. sampling rate
6. stream of pulses
7. repetition rate
8. coding process
9. analog signal
10. transmission quality
11. digital communication
12. digital transmission
13. noisy environment
14. transmission path
15. signal-to-noise ratio
16. signal levels
17. terrestrial system
18. noise power
19. binary transmission
20. reverse operation
21. 8-digit sequence
22. receiving terminal

23. frame format

24. synchronization word

二、词组翻译参考答案

1. 实现这三项功能的方案
2. 一串幅值
3. 电话质量的话路
4. 一个 8 位二进制码的序列
5. 理论上的最小抽样频率
6. 占据着 300 Hz 到 3.4 kHz 频率范围的话路
7. 每个样值 8 位码
8. 汽车点火系统的打火
9. 重复率为 64 kHz 的脉冲流
10. 真实信号与噪声信号的关系
11. 由卫星上收到的信号
12. 一条特定消息中的全部信息
13. 被传信号的波形
14. 由传输路由引入的衰减
15. 将抽样的幅值转换成一串脉冲的单元
16. 涉及第一路、第二路及其他各路的序列
17. 被称为同步字的独特的码序列
18. 地面系统
19. 脉冲的“有”或“无”
20. 高速的电子开关
21. 时分多路复用器
22. 时分多路复用

三、选择题答案

1. B 2. A 3. B 4. D 5. D 6. B

四、选择题答案

1. B 2. D 3. B 4. B 5. D 6. B 7. C 8. A 9. B

五、参考译文

1. 研究二进制的传输可见，只要简单地去判别脉冲的“有”和“无”，我们就能获得一条消息的全部信息。相比之下，许多其他形式的传输系统是利用被传信号的波形或电平来传送信息的，而这些参数又极易受到传输路径中的噪声和衰减的影响。因此选择数字传输系统在克服噪声环境的影响方面有其内在的优势。

2. 读者也许会问，解复用器怎么知道哪一组 8 位码对应于第 1 路、第 2 路及其他各路呢？显然这是很重要的。我们只要指定一个帧格式，这个问题就容易解决了。在一个帧格式里，在每一帧的开始放置一个称为帧码或同步字的独特的码序列，以标志每帧的开始，而用解复用器的一个电路去监测同步字，从而它就知道下一个 8 位码组对应于话

路1。

3. 噪声可以以多种方式被引入到传输路由里:兴许是由附近的雷击、汽车点火系统的打火,或因电信设备内部低电平的热噪声所致。正是这种被称为信噪比的东西,即真实信号与噪声信号的关系,引起了电信工程师的极大兴趣。

4. 基本的,如果信号电平比噪声电平大得多,则可产生完美的信息。但是,实际情况并非总是如此。例如,从位于遥远太空的卫星上收到的信号极为微弱,其电平仅比噪声电平稍高一点。另一类例子可在地面系统中找到,在地面系统中,尽管信号很强,噪声功率也很强。

5. 迄今为止,我们一直假定每个话路都各有一个编码器和解码器。前者是将幅度采样值转换成一组脉冲;而后者则实施反向的变换。这种设置并非必需。在实际运行的系统中,单一的编、解码器为24路、30路,甚至120个话路所共享。

6. 一个高速的电子开关被用来将每一话路的模拟信息信号依次地送到编、解码器,然后编、解码器再顺序采样幅值并把这个样值编成8位码。因此,其输出可以视为对应于第1路、第2路以及其他各路的8位码序列。

UNIT 4
WDM

一、词组翻译参考答案

1. the understanding of the properties of light
2. the fundamental importance
3. to imagine the communication system of today
4. the highways of light
5. the massive amount of information
6. to adopt new communication technologies
7. the large amount of video information
8. the wave divide multiplexing
9. to send only one wavelength
10. to transmit a large amount of wavelength
11. the error-free transmission
12. the self-healing properties
13. to access directly to the optical network
14. the video information

二、词组翻译参考答案

1. 导致WDM革命的主要进展
2. 光放大器的发明

3. 下一段光纤
4. 提高所有波长信号的功率
5. 在光放大器方面的进展
6. 增益均衡技术的发展
7. 多波长传输
8. 无线系统的增长
9. 各种应用的增长
10. 各种各样的业务
11. 处理各种业务类型
12. 全光交叉连接

三、选择题答案

1. B 2. A 3. A 4. B 5. C 6. A 7. D

四、选择题答案

1. A 2. C 3. B 4. D 5. C 6. D

五、参考译文

1. 目前全球网络业务爆炸式增长充分地体现了我们采用新的通信技术的速度。无线系统和因特网的发展就是很好的证明。

2. 近来在光网络方面最引人注目的进展出现在波分复用(WDM)方面。这些进展给地面通信系统和海底通信系统带来了好处,将可用的容量增加了几个数量级,而相应地减少了费用。

3. 然而,光放大器(OA)可以放大光纤中所有波长的信号功率,这样就不需要不同的再生器,并允许许多波长共用同样的光纤。

4. 例如,用于传输紧急电话或者实况医疗手术转播的信号与用于发送不是很紧急的可以几个小时后到达的电子邮件的信号相比,信号质量有很大的不同。

5. 进一步要考虑的是接入光网络。多数用户想要直接接入光网络以利用光网络所提供的巨大的信息容量。这将会分成几个阶段来完成。多波长光系统正快速地从核心网向终端用户扩展。

6. 波分复用是一种光的技术,它将多个波长在同一根光纤上传送,因而有效地增加了每根光纤的总带宽,该带宽可达到每个波长比特率的总和。例如,在同一光纤上有 40 个 10 Gbit/s 的单波长可将总波长提高到 400 Gbit/s,而且总带宽达到几个太比特(Tbit/s)也已成为现实。随着光电技术的进步,在同一光纤上拥有高密度波长已成为可能。因此,人们使用了“密集式波分复用”一词。DWDM 技术比 WDM 拥有更多的波长。

7. 传统的单模光纤传输 1 310 nm 和 1 550 nm 的波长,吸收了 1 340~1 440 nm 范围的波长。WDM 系统利用了在 1 310 nm 和 1 550 nm 这两个区域的波长。特殊光纤已可利用从1 310 nm到 1 600 nm 以外的整个光谱。然而,尽管新的光纤技术打开了光谱的窗口,但并非所有的光器件对整个光谱都有同样效果。例如,掺铒光纤放大器仅在 1 550 nm 附近具有最佳性能。

8. 当前,具有 16、40、80 和 128 个通道(波长)的商用系统已经问世。具有 40 个通道的系统的信道间隔为100 GHz,而具有80个通道的系统为50 GHz。具有40个通道的DWDM系统可以在一根光纤上传送总带宽400 Gbit/s。人们估计,在400 Gbit/s的速率下,传送10 000多卷的百科全书只需要一秒钟。

9. 尽管DWDM技术仍在发展,技术和标准部门也在解决着许多重要问题,但正在推出的系统仅拥有几十个波长。但是,有理由相信,在不久的将来,我们将看到在单根光纤上传送几百个波长的DWDM系统。理论上,一根光纤上可以复用1 000多个通道,200多个波长的DWDM技术已在展示之中。

UNIT 5
5G

一、词组翻译参考答案

1. higher system capacity
2. massive device connectivity
3. to connect everything
4. many other innovative technologies
5. mobile smart devices
6. to access the mobile Internet
7. the enhanced mobile broadband
8. the remote control of robots
9. massive machine type communications
10. multiple-input multiple-output
11. to over a separate frequency channel
12. a technique called beamforming
13. wearing a VR headset
14. to achieve the capacity growth
15. to be densely deployed
16. lower energy consumption
17. in the millimeter wave band
18. previous cellular networks
19. millimeter wave antennas
20. a company's virtual shop
21. to improve consumer experiences

二、词组翻译参考答案

1. 超可靠低延时服务
2. 5G规范的第一阶段

3. IMT-2020 技术的一个候选
4. 加速信息黄金时代的到来
5. 适应早期的商业部署
6. 为覆盖小区需要更多的天线
7. 用在先前蜂窝网络的大天线
8. 安装在电话杆和建筑物上的许多天线
9. 新型空间接口技术的应用
10. 与无线设备通信
11. 用于增加数据速率的另一技术
12. 组织许多天线协同工作
13. 提供足够的投资回报
14. 可用带宽
15. 大量的非结构数据
16. 产生的数据量
17. 5G 商业应用的基础
18. AR 和 VR 世界的复杂性和丰富性
19. VR 和 AR 技术的充分潜力
20. 传输几乎零延迟
21. 改善基于 AI 设备的用户体验
22. 发布第一个工业级 5G 商用芯片
23. 导致丰富的电子商务生态

三、选择题答案

1. B　2. D　3. A　4. C　5. D　6. D　7. A　8. C

四、选择题答案

1. B　2. A　3. D　4. C　5. D 6. A　7. B

五、参考译文

1. Release-16 的第二阶段将会在 2020 年 4 月完成，并提交 ITU 作为 IMT-2020 的候选技术。5G 将会连接万物，为各行各业带来好处。5G 会结合大数据、云计算、人工智能和许多其他创新技术，在下一个 10 年会加速信息黄金时代的到来。

2. 随着物联网浪潮的到来，除了智能电话外越来越多的智能设备要接入移动互联网。下一代移动通信系统，也称作第五代系统，将会提供三种场景类型：第一种是增强型移动宽带 eMBB，其目的是向以人为中心的应用提供宽带多媒体接入。

3. 为实现容量增长，5G 的小区必须密集部署，大约是 4G 网络的 40 倍到 50 倍。由于新型空间接口技术，以及各种各样服务和终端的应用，一个典型的 5G 节点需要配置大约 2 000 个参数。5G 规划还旨在降低延迟和降低能耗，以更好地实施物联网(IoT)。

4. 5G 网络通过使用更高的无线电波来实现更高的数据速率，毫米波频段在28 GHz 至 39 GHz 之间，而之前的蜂窝网络使用的频率范围在 700 MHz 至 3 GHz 之间的微波频段。一些供应商会使用微波频段的第二低的频率范围，低于 6 GHz，但这将不会有新频率的高速率。由于这些频率的带宽更丰富,5G网络将会使用更宽的频率通道与无线设备通信,最高可达400 MHz,而4G LTE为20 MHz,每秒可传输更多的数据比特。

5. 毫米波被大气中的气体吸收,比微波的作用范围小,因此小区的覆盖范围受到限制;5G的小区范围有街区那么大,而以前的蜂窝网络可能有好几英里宽。电波穿透建筑物的墙壁也有困难,需要多个天线来覆盖小区。

6. 另一种提高数据速率的技术是大规模MIMO(多输入多输出)。每个单元将有多个天线与无线设备通信,每个天线通过一个单独的频率通道,由设备中的多个天线接收,因此多个比特流数据可以同时并行传输。在一种称为波束形成的技术中,基站计算机将不断计算无线电波到达每个无线设备的最佳路径,并将组织多个天线作为相控阵列一起工作,产生毫米波波束到达设备。

7. 物联网是5G商用的基础,发展5G是为了给我们的生产和生活带来便利。而物联网就为5G提供了一个大展拳脚的舞台。在这个舞台上5G可以通过众多的物联网应用,如智慧农业、智慧物流、智能家居、车联网、智慧城市等,发挥出自己强大的作用。

8. AR和VR世界的复杂性和丰富性需要处理大量数据。当前的4G网络标准受到一些限制,例如与带宽、等待时间和一致性有关的限制,特别是当需要远程馈送数据时。在这方面,5G可能会释放VR和AR技术的全部潜力。5G更快的速度和更低的延迟将有助于克服这些弱点。

UNIT 6
The Vision of 6G

一、词组翻译参考答案

1. the 5G base station
2. the future 6G network services
3. enhanced mobile broadband
4. to guarantee network coverage
5. the unmanned aerial vehicle
6. the low power consumption
7. security and privacy
8. at the level of terahertz
9. beamforming
10. to increase channel capacity
11. self-driving cars
12. the improved driving safety
13. the location services
14. the radio access network
15. the high-resolution imaging
16. to improve spectrum utilization
17. the optimization of radio interface
18. the greater propagation loss
19. a service-based architecture
20. the sensing and imaging equipment
21. the edge cloud of the future
22. to realize accurate positioning

二、词组翻译参考答案

1. 消费者对业务日益增长的需要
2. 对垂直行业生产力的要求

3. 商务和社会的需要
4. 数千的无线连接
5. 开放太赫兹频谱
6. 6G 技术研究和发展的促进
7. 6G 支持的应用场景
8. 移动通信性能
9. 高水平安全和隐私
10. 全息和触觉互联网
11. 频谱效率的改进
12. 高速大容量通信
13. 高级虚拟/增强型用户体验
14. 新的网络架构
15. 网络拓扑的动态性
16. 基于先进 AI 的完全定制化
17. 满足未来的挑战
18. 空间复用技术
19. 多 MIMO 天线配置
20. 网络的不同层级
21. 有效的云实现
22. 引入附加的网络功能
23. 网络技术的快速发展
24. 人们的主要娱乐工具
25. 传输不同地方的人的三维全息图

三、选择题答案

1. C　2. A　3. A　4. D　5. B　6. C　7. B

四、选择题答案

1. A　2. D　3. B　4. A　5. C　6. B　7. C

五、参考译文

1. 5G 的驱动力来源于消费者不断增长的流量需求，以及垂直行业的生产力需求。也就是说，本质上是商业需求驱动了 5G 的发展。

2. 11 月 3 日，中国宣布成立国家 6G 技术研发推进工作组和总体专家组，这标志着我国 6G 技术研发工作正式启动。6G 通信技术支持五种应用场景。

3. 综上所述，6G 的主要需求是：高速大容量通信、极速扩展覆盖、低功耗、低延迟、高可靠通信、海量连通性和传感、高安全性和私密性。

4. 另一方面，尽管鉴于 5G 技术已经取得的成就，6G 的频谱效率不会有大的提升，但高速、大容量的通信将是提供先进的虚拟/增强用户体验的基础。

5. 6G 可以考虑使用多种技术来应对未来的挑战，其中许多技术适合在研究协作中进行联合评估。在这里，我们强调了一些技术组成，我们认为这些技术是我们预计 6G 将具有的新功能的潜在技术成因。

6. 6G信号的频率已经在太赫兹级别，而这个频率已经接近分子转动能级的光谱了，很容易被空气中的被水分子吸收，所以在空间中传播的距离不像5G信号那么远，因此6G需要更多的基站“接力”。

7. 6G将使用“空间复用技术”，6G基站将可同时接入数百个甚至数千个无线连接，其容量将可达到5G基站的1 000倍。更高的频率意味着更高的传输损耗。

8. 随着5G发展，未来的移动通信网络将进入各行各业。大部分的垂直行业都需要定位服务，比如资产跟踪、精准营销、运输和物流、AR、医疗保健等应用，但传统卫星定位方法在城市和室内场景中并不精准。

UNIT 7
Cloud Computing

一、词组翻译参考答案

1. cloud computing
2. business computing model
3. all kinds of software service
4. virtual computation resources
5. large scale data center
6. a better use of distributed resources
7. super computing power
8. to reduce the overall cost
9. with the help of a browser
10. ubiquitous network access
11. the definition of cloud computing
12. the principal service mode
13. the paramount example of cloud computing
14. layers of cloud computing architecture
15. to provide a computing platform
16. computing platform
17. the cloud infrastructure
18. basic computing resources

二、词组翻译参考答案

1. 获得计算能力
2. 这种资源池
3. 一些大规模服务器群

4. 通过软件自动管理

5. 使用云计算的用户

6. 世界上最大的数据中心

7. 主要的云供应商

8. 随着互联网的大规模普及

9. 解决大规模计算问题

10. 借助浏览器

11. 下一代计算模型

12. 可配置计算资源共享池

13. 通过互联网接入在线资源

14. 在用户系统上安装和运行应用程序

15. 使用托管在云服务器上的服务

16. 软件生命周期

17. 获得快得多的服务

三、选择题答案

1. A　2. D　3. B　4. B　5. A　6. C

四、选择题答案

1. A　2. C　3. D　4. B　5. A　6. C　7. A

五、参考译文

1. 云计算是一种新出现的商业模型。它将计算任务分配给由大量计算机组成的资源池,并根据需要使每种应用系统获得计算能力、存储空间和各种软件服务。这种资源池被称为“云”。

2. 云计算会集中所有计算资源并能够通过软件不间断地自动管理。使用云计算的用户不必担心乏味的细节并可以集中精力于自己的事业。云计算是信息技术的新趋势,它将计算及数据从桌面和手提电脑上转到大型数据中心。

3. 像谷歌、微软和亚马逊这样大的云计算提供商在美国和世界其他地方已经建立了世界上最大的数据中心。每个数据中心包括成百上千的计算机服务器、冷却设备和变电站变压器。

4. 在云计算服务中,数千的计算机构成超级服务器,由这些服务器向用户提供强有力的数据计算和处理能力,这些能力是个人计算机难以实现的。这一结果会降低整个费用。在云计算中,用户借助浏览器接入数据、应用或任何服务而不管所使用的设备和用户的位置。

5. 随着互联网在全世界的大规模普及,应用业务可以通过互联网来提供。云计算的主要目标是更好地利用和整合分布式资源以获得更高的数据吞吐量和解决大规模计算问题。云计算具有超强的计算能力。

6. 云计算因其按需自我服务、无所不在的网络接入、与位置无关的资源池以及风险转移的主要优势而被看成是下一代计算模式。云计算是继分布式计算、并行计算和网格计算后最新发展的计算模式。

7. 云计算就是互联网计算,通常互联网被看成是云的集合,这样云计算可以定义为使用互联网向个人和组织机构提供服务的技术。云计算使得顾客能够随时随地通过互联网接入在线资源,而不必担心最初资源的管理和维护。

8. 云应用通过互联网提供软件即服务(SaaS),这样用户就不必在其用户系统上安装和运行应用软件了。根据客户的需要,软件作为一种服务提供给客户,使用户能够使用托管在云服务器的服务。谷歌应用是最普遍使用的 SaaS。

9. 客户不必购买所需要的服务器、数据中心或网络资源。其主要的优势在于客户只需对使用服务的这段时间付费。其结果是客户用很少的成本获得快得多的服务。

UNIT 8
Edge Computing

一、词组翻译参考答案

1. the high risk of private data leakage
2. the smart home applications
3. the data privacy preservation
4. to provide prompt responsiveness
5. the cloud computing data centers
6. the key factors for system reliability
7. the data privacy and security
8. to collect data from sensors
9. in edge computing environment
10. to eliminate the processing workload
11. to provide lower latency
12. the deployment of various IoT devices
13. to urge the creation of edge computing
14. the huge volume of data transportation
15. the edge computing
16. the cloud computing paradigm
17. a service provisioning model
18. the scalable distributed capabilities

二、词组翻译参考答案

1. 无法计算和存储
2. 智能家居的完美替代者
3. 最先进的云计算中心
4. 卸载到网络边缘

5. 减少云数据中心的压力
6. 最重要的应用场景
7. 利用云计算的好处
8. 当今互联网的主导地位
9. 各种应用产生的数据量
10. 将收集的数据发送到远程云数据中心
11. 为边缘数据处理提供完美的平台
12. 定制应用和访问云服务
13. 成本优化和增加的竞争力
14. 产生大量的公共业务相关数据
15. 提供处理或存储关键数据的能力
16. 日益增加的对实时数据处理的商业需求
17. 对服务质量和实时响应的严格要求

三、选择题答案

1. B　2. D　3. A　4. C　5. D　6. A　7. B

四、选择题答案

1. B　2. D　3. A　4. C　5. A　6. D　7. C　8. B

五、参考译文

1. 云服务提供商(CSPs)通过提供软件即服务(SaaS)、平台即服务(PaaS)和基础设施即服务(IaaS)等服务,为最终用户提供业务灵活性和高效率。例如,服务供应商可以扩展服务来满足他们的需求,定制应用程序,并在任何有互联网连接的地方访问云服务。因此,对于互联网带宽需求有增长或变动的企业来说,基于云的服务是理想的选择。

2. 尽管云计算可以为企业提供动态的、基于云的运营模型,以实现成本优化和提高竞争力,但在工业物联网、联网自动车辆(CAVs)、智能家居和智能城市等许多场景中,它也有一些缺点。例如,基于云计算的处理需要从终端设备和传感器传输大量数据,这会消耗大量网络带宽。

3. 随着物联网、5G通信、自动驾驶和智慧城市的发展,边缘计算正在连接和弥合众多终端设备和集中式云计算数据中心之间的差距。此外,在数据隐私和安全是主要问题的情况下,边缘计算承诺提供数据隐私保护是通过将数据保持在网络边缘,而不是将数据发送到云数据中心来实现的,进而提供更低的延迟,增加可靠性,提高整体网络效率。

4. 由于数百万的边缘设备以分布的方式进行地理部署,以及处理数据也是在异构分布式设备上执行的,所以设计适合边缘计算环境的新型系统结构非常重要。运行在边缘计算环境中的各种应用程序和设备产生的数据是大量和高度异构的。

5. 在家中部署各种传感器并将收集到的数据发送到远程云数据中心进行处理,会给大量本地居民带来私人数据泄露、数据滥用和物理威胁的高风险。因此,传统的基于云计算的数据处理不适合于智能家居应用,数据隐私保护的边缘计算成为智能家居的完美替代品。

6. 然而,由于一个典型的城市也会产生大量的公共服务相关数据,即使是最先进的云数据中心也无法实时处理这些数据,以进行城市规模的交互式分析,这是由于缺乏计算、存储和联网的能力。如果数据处理可以被转移到网络的边缘,它可以减少云数据中心的压力,使近乎实时的分析成为可能。

7. 虽然传统的云计算技术在实时响应、隐私保护、低能耗等方面无法满足需求,但边缘计算范式在本质上并不是要取代云计算技术。相比之下,云计算和边缘计算在许多场景中是互补的、相辅相成的。

8. 边缘计算技术可以充分利用边缘设备的计算能力,在边缘设备上进行部分或整体计算,从而降低云数据中心的计算需求和核心网络的传输带宽。边缘计算和云计算的合作为物联网无所不在的数据分析和延迟关键应用(如自动驾驶和工业网络系统)的低延迟计算提供了更多的机会。

UNIT 9
The Development and Application of IoT

一、词组翻译参考答案

1. the vision of the IoT
2. GPS and laser scanner
3. the space coupling
4. orientation tracing
5. the voice control
6. the infrared sensor
7. the alert system
8. the hearing impaired user
9. the smart sensor devices
10. the concept of IoT
11. the global positioning system
12. the function of recognition
13. be connected via the IoT
14. these autonomous vehicles
15. wireless sensor network
16. the sight and mobility limitation
17. the additional safety features
18. the notion of driverless cars
19. total smart products consumption
20. the concept of home automation

二、词组翻译参考答案

1. 连接虚拟世界与物理世界

2. 合并许多不同的部件
3. 许多行业和应用
4. 另一个科技和经济浪潮
5. 全球信息业
6. 国家信息战略
7. 实现智能识别
8. 信息外部设备
9. 射频识别设备
10. 一种主动诊断技术
11. 实现非接触信息传输
12. 射频信号发射单元
13. 数据采集和后端数据库
14. 电子标签和阅读器接口
15. 大量低成本传感器节点
16. 与真实世界交互
17. 嵌入式芯片、传感器和其他智能物体
18. 聚集并利用集体智能
19. 工业物联网术语
20. 第四次工业革命
21. 控制智能设备和应用
22. 连接不同的家庭智能产品
23. 对新产品的高适应性

三、选择题答案

1. A　2. D　3. C　4. B　5. C　6. A

四、选择题答案

1. B　2. D　3. A　4 .D　5. C　6. B　7. C

五、参考译文

1. 当司机驾驶汽车操作有误时，汽车会自动告警；公文包会提醒物主忘了拿什么东西；衣服会告诉洗衣机衣服的颜色及需要的水温等。

2. 物联网开始于 1999 年，其最初的设想是要连接虚拟世界与物理世界。然而，从早期一出现其发展就引人注目，现在很多不同行业都有其应用，它已经成为当下社会日常生活中非常重要的一部分。

3. 物联网被看作是继互联网之后全球信息业的另一个科技和经济浪潮，吸引了政府、企业和学术界的高度关注。美国、欧洲和日本甚至将其列为国家信息战略。

4. 物联网的定义是：通过射频识别 RFID、红外感应器、全球定位系统 GPS、激光扫描器等信息传感设备，按照约定的协议，把任何物品与互联网连接，进行信息交换和通信，以实现智能化识别、定向追踪、监控和管理。

5. 在 IoT 中的“物”不仅要实现连在一起，也要实现识别、定位、追踪、管理等功能，这

就要求能识别所有的物品。射频识别 RFID 技术就是所有物品连接的“说话技术”,因此 RFID 在这些关键技术中具有突出的重要地位。

6. RFID 系统由数据获取和后端数据库网络应用组成。已经发布或正在开发的标准主要涉及数据采集,包括电子标签和阅读器接口、阅读器和计算机之间的数据交换协议、RFID 标签和阅读器的性能、RFID 标签数据编码标准内容。

7. 在政府倡导的工业化和信息化结合的背景下,IoT 将是工业和其他信息业的真实启动点。一旦 IoT 大规模流行开来,更多的小物体需要安装智能传感器。用于动物、植物、机械和其他物品的传感器和电子标签将会大大超过现有电话的数量。

8. 另一有可能被彻底改变的重要行业是交通运输业。还在5年前,我们多数人还认为无人驾驶汽车的概念是科幻小说,或者至少是几十年后的事。

UNIT 10
Artificial Intelligence

一、词组翻译参考答案

1. artificial intelligence
2. the field of computer science
3. to mimics “cognitive” functions
4. the commercial potential of AI
5. such as expert systems
6. the natural language processing
7. to mimic the problem-solving of humans
8. ML programs
9. the active area of research
10. to learn from data
11. the deep learning
12. a specific type of machine learning
13. higher-level abstract representations
14. a set of machine-learning algorithms

二、词组翻译参考答案

1. 由机器表现出的智能
2. 相比于自然智能
3. 系统正确解释外部数据的能力
4. 实现特定目标和任务
5. AI 在一些领域潜在的商业影响
6. 语音识别和自然语言处理

7. 不同AI分支的任务
8. 要求模式识别、推理和决策
9. 远超博弈的影响
10. 由于专家系统的流行
11. 当时令人惊讶的程序
12. 模拟人类专家知识
13. 使用在数据挖掘和医疗诊断
14. 想出明确的规则
15. 如图像分类、语音识别或语言翻译
16. 解决明确定义的逻辑问题

三、选择题答案

1. A 2. C 3. D 4. B 5. B 6. D 7. A

四、选择题答案

1. A 2. D 3. B 4. C 5. B 6. A

五、参考译文

1. Kaplan 和 Haenlein 将 AI 定义为“一个系统正确解释外部数据的能力,从这些数据中学习的能力,以及通过灵活的适应来利用这些学习以实现特定目标和任务的能力”。通俗地说,“人工智能”一词是指机器模仿人类与其他人类思维相关的“认知”功能,比如“学习”和“解决问题”。

2. 例如麦肯锡公司预测 AI 在一些领域潜在的商业影响市场价值上万亿美元。所有这些都是由于 AI 在过去十几年突然取得的令人惊讶的进步所带来的。阿尔法狗、自动驾驶汽车、Alexa、Watson 以及其他类似的系统,在博弈、机器人、计算机视觉、语音识别和自然语言处理等方面都取得了令人惊异的进步。

3. AI 所蕴含的最终能力可能会终结人类所做的所有事情,并且会比人类做得更好,也就是说能实现超人的性能,就像近来阿尔法狗和阿尔法零所见证的那样。这一含义有时被称为人工智能的中心法则。从历史发展来看,术语 AI 集中反映在下列方面。

4. AI 不同分支的许多任务有共同的特性。他们都需要复杂条件下的模式识别、推理和决策。而且,他们经常要处理不明确的问题、噪声数据、模式不确定性、组合上的大搜索空间、非线性以及快速解决方案的需要。

5. 回望 AI 的 30 年历程,可以确认三件里程碑式的早期事件。第一件是深蓝在 1997 年国际象棋比赛中击败世界冠军卡斯帕罗夫,第二件是 IBM 超级计算机 Watson 在 2011 年成为美国游戏节目 Jeopardy 的冠军,第三件是 2016 年阿尔法狗在围棋比赛中令人惊讶的获胜。人工智能的发展使这些惊人的壮举成为可能,现在它的影响已经远远超出了游戏的范畴。

6. 虽然启发式搜索和专家系统适合解决如下棋这样明确定义的逻辑问题,但是却很难处理更复杂和模糊的问题,如图像分类、语音识别或者语言翻译。于是出现了 AI 的另一个繁荣:机器学习 ML。

7. ML 不是一个新概念。在 1959 年,ML 的先驱者之一 Arthur Samuel 将机器学习定义为:使计算机不必经过明确编程就能够学习的研究领域。也就是,ML程序(像if-then语句)没有明确地输入计算机。某种意义上,ML程序面对数据时能够相应地调整自己。

8. 人工智能是一个广泛而活跃的研究领域,但它不再是学术界的唯一领域,越来越多的公司将人工智能应用到他们的产品中。谷歌一直是使用机器学习的先锋,与盲目地按照指令学习不同,机器学习是一种能够向数据学习的计算机系统。特别是公司使用了一些称作深度学习的机器学习算法,这些算法允许计算机从大量的数据中进行模式识别。

UNIT 11
IP Version 6

一、词组翻译参考答案

1. to grow at a dramatic rate
2. the current version of the IP protocol
3. to include two protocol stacks
4. the base IPv6 header
5. the dual stack mechanism
6. to identify a flow
7. the adequate IP addresses
8. the type of the header
9. to fill the fields in the header
10. an overabundance of addresses
11. the basic addressing support
12. to solve the address problem
13. the existing advantages of IPv4
14. the IPv4's processing performance
15. to define standards for hosts
16. a contiguous string of zeros

二、词组翻译参考答案

1. 地址自动配置
2. 对集中服务器的需要
3. 支持移动IP协议
4. 定义明确的消息
5. 源和目的IP地址
6. 表示协议版本
7. IP协议的基本字段

8. 满足互联网增长的需要
9. 提供安全和移动性的更好支持
10. 下一代互联网的基础
11. 提供 IPv6 报头的描述
12. 大地址空间的优势
13. 网络发展必然的趋势
14. 马上将其网络改变成 IPv6
15. 为实现平稳和逐步的转换
16. 实现地址和协议转换
17. 对应整个 16 位地址的十六进制值

三、选择题答案

1. A 2. D 3. B 4. C 5. C 6. B

四、选择题答案

1. B 2. D 3. A 4. C 5. A 6. D 7. B

五、参考译文

1. 20 世纪 80 年代末期和 90 年代初期,互联网开始以惊人的速度发展,工程师们意识到当前 IP 协议的版本不足以满足互联网发展的需求。特别令人关切的是为访问 Internet 的各种设备提供足够的 IP 地址可用性。

2. 根据 Gartner 发布的报告,到 2020 年将有 250 亿个物联网设备出现,大量的感知终端或机器接入移动网或互联网,必将引发对 IP 地址的巨大需求,而 IPv6 是解决地址问题的根本途径。

3. IPv6 还提供了额外的一些功能。地址自动配置允许主机自动配置他们的 IP 地址,而不需要一个中央服务器。为了更高效地处理,包含 IP 协议工作的基本字段的报头已被简化;即使是 40 字节,IPv6 报头也比 IPv4 报头更适合报头压缩。

4. 随着互联网的飞速发展,IPv4 协议已经不能满足用户的需求。这主要是由于 IPv4 在地址、路由和安全性方面的限制。相应地,IPv6 在大地址空间、安全性、移动性、服务质量等方面具有优势。因此 IPv6 协议已成为网络发展的必然趋势。

5. 有了基本的寻址支持和额外的支持,特别是安全性和移动性,新的 IPv6 协议可以形成下一代互联网的基础。因此,理解它的一些基本操作很重要。互联网协议通常为主机和路由器的通信定义标准。IPv6 就是这样一个重要的协议。

6. 为 IP 软件携带的信息通常被称为“报头”,它指定了必须如何解释和处理消息。报头的定义和语义必须明确,以便发送方和接收方能够通信。报头“格式”就是这样做的。

7. 源代码可以标记这些位来请求某些“差异化服务”,路由器可以提供相应的转发行为。这些位可能被中间路由器再次标记,以确保流符合所做的约定。字段的其余两位保留为“显示拥塞通知(ECN)”,用于预先通知传输协议数据包所走路径上的拥塞情况。

8. IPv6 报头后面的包的其余长度由有效负载长度表示。IPv6 报头之后的报头类型由下一个报头标识。对于数据而言,这通常是 TCP 或不可靠数据报协议 UDP。然而,IPv6 定义了可能存在的多个扩展头。

UNIT 12
Circuit Switching and Packet Switching

一、词组翻译参考答案

1. circuit switching
2. packet switching
3. message switching
4. subnet
5. header
6. destination address
7. error control
8. store-and-forward manner
9. bursty
10. transmission delay
11. intermediate switching equipment
12. switching technique
13. return signal
14. message processor
15. given maximum length
16. information transfer
17. random
18. dedicated circuit
19. channel utilization

二、词组翻译参考答案

1. 存储和处理用户数据的能力
2. 特定的信令信息
3. 被精心定义的称为报文的数据块
4. 涉及源和目的地址的信息
5. 称为报文处理器的计算机
6. 存储转发传输技术
7. 带宽的动态分配
8. 报文的整个传输时延
9. 交换技术
10. 电路交换
11. 报文交换

12. 分组交换
13. 连线的整个通路
14. 源到目的地的一对
15. 通信各方
16. 传输单元
17. 在建立电路时产生的起初连接成本
18. 用户所需的短时延的限制
19. 固定专用的端到端电路
20. 低的电路利用率

三、选择题答案

1. B　2. A　3. C　4. A

四、选择题答案

1. C　2. B　3. C　4. B　5. D

五、参考译文

1. 在电路交换中，当呼叫发生时，从呼叫源点到终点之间要建立整个通路的连接，而且在通信双方释放该电路之前，此通路一直保持分配给这对源点-终点(不管通路是否使用)。被称为电路交换机的交换设备没有存储或控制用户送往终点路由的数据的能力。电路由特殊的信令来建立，该信令提供网络选择路由，并在其进程中确定信道。一旦电路建立，就由一个返回信号通知呼叫源开始传输。

2. 在报文交换中，传输单元是一个被精心定义的数据块，该数据块称为报文。除了要传的内容外，报文还包括有报头和检验项。报头含有源地址和目的地地址的信息，还有其他的控制信息；检验项用于误码控制。交换单元是一个被称为报文处理器的计算机，它具有处理和存储的能力。

3. 报文独立并异步地传输，在源点和终点间选择自己的传送路由。首先报文由主机送到与它相连的报文处理机。一旦报文被完全收到，报文处理机就检查其报头，并相应地决定该报文传送的下一个输出信道。如果这个所选信道忙，则该报文就排队等待，直到该信道空闲时开始发送。

4. 分组交换是报文交换的一种变形。在分组交换中，报文按指定的最大长度分成若干个被称为分组的段。与报文交换一样，每个分组都含有一个报头和检验项。分组以存储转发的方式独立传送。分组交换具有报文交换所具有的优势并另具特色。

5. 在电路交换的情况下，建立电路总要对开始的接续付出代价。只有在这种情况下，即一旦电路建立，有持续不断的信息流传送，以便分担初始花费，才能提高价格效率比。传统方式的话音通信正是这种情况，而电路交换也确实是电话系统中使用的技术。

6. 但是，计算机间的通信具有突发性。突发是由报文产生过程和报文长度方面高度的随机性造成的，也是用户对时延的要求很短所造成的。用户和设备不常使用通信资源，但当他们使用时，他们就要求其具有相对迅速的反应。

7. 因此，对突发性的用户，存储转发传输技术提供了一个更低价高效的解决办法，因为只有在报文传送的时间里，报文才占据着一条特定的链路。其他时间，它被存储在某个中间交换机中,此时的链路又可用于其他的传输。这样与电路交换相比,存储转发方式的主要优点是对通信带宽的动态分配,并且这种分配是以网络中的特定链路和特定报文为基础的。

8. 除了以上讨论的优点外,分组交换还具有一些特点。它提供动态分配带宽的全部优势,甚至在报文很长的时候依然如此。的确,在分组交换的情况下,一个报文的许多分组有可能在从源点到终点的通路中的多条链路上同时传送着,因而达到"管道传送"的效应。与报文交换相比,它大大地减少了整个报文的传送时延。

9. 除了以上讨论的优点外,分组交换还具有一些特点。在中间交换设备中,这种方式只需要较小的存储分配区域。分组交换的误码特性较好,由于它只涉及很短的长度,因而导致了更高效的纠错方式。当然,分组交换亦有自己设计中的麻烦,例如,当报文无序地到达目的节点时,需要重新对该报文进行分组的排序。

UNIT 13
EPON

一、词组翻译参考答案

1. the mature Ethernet technology
2. the ideal access method
3. the backbone network
4. the copper line
5. the dial up line
6. to provide enough bandwidth
7. the video conference
8. the next generation access network
9. the technology of user access network
10. the optical line terminal
11. the optical network unit
12. the multi-service platform
13. to perform bandwidth assignment
14. to adopt the broadcast method
15. to extract its own data frame
16. to separate the ONU's upstream channels
17. the optical splitter
18. to provide optical interface
19. to prevent data collision
20. the main advantage

21. the assigned time slot

二、词组翻译参考答案

1. 互联网的发展
2. 接入网的瓶颈
3. 更大的容量
4. 高速局域网
5. 提供足够的带宽
6. 互动游戏
7. 数据和视频业务
8. 用户接入网技术
9. 10 Mbit/s 的接入带宽
10. 高质量互联网业务
11. 城域网
12. 多业务平台
13. 光分配网络
14. 点到多点网络
15. 从不同 ONU 来的数据流
16. 避免数据冲突
17. 使用波分复用

三、选择题答案

1. A　2. C　3. D　4. B　5. A　6. C　7. D

四、选择题答案

1. A　2. C　3. B　4. C　5. C　6. C　7. A

五、参考译文

1. 今天，随着互联网的快速发展，越来越多的互联网业务逐步走进千家万户，接入的瓶颈效应开始浮现。EPON 技术，结合了成熟的以太网技术和高带宽 PON 技术，是综合业务理想的接入方式。

2. 互联网业务的剧增暴露了接入网容量的不足。被称为"最后一公里"的接入网仍旧是高速局域网和大容量骨干网间的瓶颈。

3. 此时需要一种新型的接入技术，它应有如下特性：便宜、升级简单并同时提供声音、数据和视频业务。EPON 技术，结合了低费用以太网技术和低费用光网络技术，是面向未来的下一代接入网技术的最佳代表。

4. 以太无源光网络(EPON)是下一代用户接入网络技术，它在光线路终端(OLT)和许多光网络单元(ONU)或光网络终端(ONT)之间铺设光纤接入线。光线路终端(OLT)位于网络侧，而光网络单元或光网络终端(ONU/ONT)在用户侧。

5. 除了网络汇集和接入功能外，OLT 也可以根据用户的不同服务质量需求完成带宽分配、网络安全和管理配置。ONU(光网络单元)位于端用户或路边，可以提供连接到 OLT 的光接口以及用户端如话音、数据和视频的业务接口。

6. 然而,从不同ONU同时传输的数据流仍会产生冲突。这样,在上行方向(从用户到网络),PON应该采用一些信道分离机制以避免数据冲突并公平地共享主干光纤信道容量和资源。

7. 在时分复用(TDM)PON技术中,当从几个ONU同时传输的数据到达组合器时会产生冲突。为避免冲突,每个ONU必须在其自己的传输窗口(时隙)进行传输。TDM PON的主要优点是所有的ONU都工作在同样的波长。

8. 很有可能,EPON和点到点光以太网为电信部门的转向提供最佳的可能性。随着业务供应商投资于光接入技术,新应用将成为可能,从而刺激收入增长并驱动骨干网上更多的业务量。

UNIT 14
IPTV

一、词组翻译参考答案

1. the communications industry
2. the broadcast television
3. the infrastructure
4. the video telephone
5. the broadband digital technology
6. the end-to-end transmission
7. the switch cabinet
8. the high-definition television
9. the coding technologies
10. the International Telecommunication Union
11. the content provider
12. the interactive link

二、词组翻译参考答案

1. 通过IP网络传送电视
2. 多种多样的内容和通信服务
3. 可视电话和远程监控
4. 利用电视屏幕向用户发送数字电视频道
5. 例如话音连接的电信服务
6. 有限的网络带宽
7. 先进的编码技术
8. 国际电信联盟
9. 大量的创新业务和用户应用

三、选择题答案

1. B 2. D 3. D 4. A 5. B

四、选择题答案

1. C 2. A 3. B 4. A 5. D

五、参考译文

1. 简单来说,IPTV 意味着通过 IP 网络提供电视节目。对于电信公司来说,IPTV 是一种全新的多媒体体验,它扩展了传统广播电视的范围;IPTV 是一种综合的、"包罗万象的"媒体平台,可以从单一业务供应商通过单一网络向单一用户设备提供多种服务内容和通信服务。

2. 为什么使用 IP 技术来传输视频?答案是:IP 是孕育成功的奇妙平台,IP 已经无孔不入。第二个动机是 IP 能够提供比传统电视更多的功能。IP 的开放和扩展特性使其有希望在未来提供更多的服务。IP 的体系结构也可以用来提供视频电话和远程监控,以及新的视频应用。

3. 关于 IPTV 的含义有多种观点。对于广播业务服务商来说,IPTV 仅仅是一种新出现的使用 TV 屏幕向家庭用户发布数字电视频道的平台。对于电信业来说,IPTV 与提供话音、数据和视频的新的宽带数字技术同义。IPTV 是对现有卫星、有线系统和地面系统的补充,虽然有时候是现有系统有力的竞争者。

4. 迄今为止,电信行业一直只是提供如两点间话音通信的电信业务。电信公司并不关心所承载的信息内容。当电信公司面临其传统话音和宽带通信日益减少的用户收入时,他们开始涉足 IPTV。

5. IPTV 可以潜在地向用户提供大量的新型业务,许多这类新业务已经在数字电视中成为可能,主要的区别在于 IPTV 使用双向通信通道,所以用户可以与业务内容和业务供应商直接进行交互。供应商和用户间的交互链路能够按用户的请求向用户家中的单一设备发送单一视频流,相比来说,广播方式是所有信道同时发送给所有用户。

6. 虽然技术问题并不简单,但也不是不可逾越的。然而,法律和规则问题才真正让人头疼。一个突出的问题是确定哪一个国家管理机构负责 IPTV 业务的管理。这取决于对 IPTV 的确切定义。一些国家认为 IPTV 是广播业务,而另一些国家认为它是电信业务。不但国家的司法体制不同,而且其电信公司和广播业务服务商也不同。

UNIT 15
BlockChain

一、词组翻译参考答案

1. no access restrictions
2. a term used in information technology
3. a shared database

4. a reliable cooperation mechanism
5. a broad application prospect
6. the distributed ledger
7. the data in any given block
8. the ledger of the crytocurrency bitcoin
9. to restrict users' reading rights
10. the first digital currency
11. the document timestamp
12. to verify and audit transactions
13. to compels offer and acceptance
14. the data stored on the blockchain
15. a cryptographic hash of previous block
16. the execution of a consensus protocol

二、词组翻译参考答案

1. 不用修改随后所有数据块
2. 作为公共交易账簿
3. 由网络管理员邀请的
4. 由公共网络提供的管制水平
5. 记录许多计算机上的交易
6. 存在已久的再次支付问题
7. 缺少集中的脆弱点
8. 将数据批处理为带时间戳的块
9. 用区块链作为传输层
10. 使用区块链技术来记录交易
11. 基于区块链的智能合约
12. 智能合约的主要目标
13. 使用P2P网络和分布式时间戳服务器
14. 允许将多个文档证书收集到一个块中
15. 将区块链纳入其会计和记录程序
16. 加密货币比特币的核心部件
17. 网络上所有交易的公共账簿
18. 比特币区块链和以太坊区块链
19. 将敏感数据暴露给公共互联网的风险

三、选择题答案

1. C　2. D　3. A　4. B　5. B　6. C　7. D

四、选择题答案

1. B　2. D　3 .C　4. A　5. A　6. C　7. B

五、参考译文

1. 区块链是一个信息技术领域的术语。从本质上讲，它是一个共享数据库，存储于其中的数据或信息，具有不可伪造、全程留痕、可以追溯、公开透明、集体维护等特征。基于这些特征，区块链技术奠定了坚实的“信任”基础，创造了可靠的“合作”机制，具有广阔的运用前景。

2. 一旦被记录下来，任何给定块中的数据都不能在不改变所有后续块的情况下进行回溯性修改，这需要网络多数人的一致同意。虽然区块链记录不是不可改变的，但区块链的设计可以被认为是安全的。因此分散的共识被认为是一个区块链。

3. 比特币的设计启发了其他应用，公众可读的区块链被加密货币广泛使用。区块链被认为是一种支付方式。私有区块链已经被提议用于商业用途。来自《计算机世界》的消息称，这种没有适当安全模型的区块链营销是“蛇油”。

4. 中本聪在一个重要方面改进了设计，使用类似 Hashcash 的方法将块添加到链中，而不需要由信任的一方签名。该设计于 2009 年由中本聪实施，作为加密货币比特币的核心组成部分。在比特币中，它是网络上所有交易的公共总账。

5. 公共区块链绝对没有访问限制。任何有互联网连接的人都可以向它发送事务，同时也可以成为一个验证器(参与协商一致协议的执行)。一些最大、最知名的公共区块链是比特币区块链和以太坊区块链。私有区块链要得到许可才能加入。

6. 开放的区块链比一些传统的所有权记录更容易使用，这些记录虽然对公众开放，但仍然需要物理访问才能查看。因为所有早期的区块链都是无须许可的，所以关于区块链定义的争论就产生了。在这场正在进行的争论中，有一个问题是，一个拥有由中央授权的验证者的私有系统是否应该被认为是一个区块链。

7. 区块链技术可以集成到多个领域。如今，区块链的主要用途是作为加密货币的分布式账本，最显著的就是比特币。大多数加密货币使用区块链技术来记录交易。例如，比特币网络和以太坊网络都是基于区块链的。基于区块链的智能合约是一种建议的合约，它可以部分或全部执行，也可以在没有人工交互的情况下强制执行。

UNIT 16
Software-Defined Networking

一、词组翻译参考答案

1. the advent of cloud computing
2. to simplify network management
3. the upper layer of SDN architecture
4. an application layer that defines rules
5. an abstraction of the network topology
6. the flow tables and data handling policies

7. to reduce both CapEX and OpEX
8. many technological hurdles
9. high availability requirements
10. to guarantee easy management
11. inherited from existing adopted technologies
12. the network control and the forwarding process
13. this network segmentation
14. the network flexibility and controllability
15. the system virtualization and cloud computing
16. the complexity of network elements
17. the frequent network failures
18. the routing of traffic

二、词组翻译参考答案

1. 消除为了维护过程
2. 从网络灵活性和可控性来说
3. 软件定义网络的出现
4. 虚拟化和云计算的优势
5. 为了便于网络管理
6. 网络单元的多样性和复杂性
7. 在经常的网络故障下
8. 分开路由选择和转发判决
9. 有关逻辑网络拓扑的信息
10. 在控制平面建立的配置
11. 决定网络策略的控制器
12. 通过应用层分配资源到用户
13. 在数据平面设立物理网络单元
14. 与现有协议的后向兼容性
15. 通过南向API收集网络信息
16. 提供管理业务量的标准化方式

三、选择题答案

1. A 2. D 3. B 4. C 5. D 6. B

四、选择题答案

1. A 2. D 3. B 4. C 5. B 6. A 7. D

五、参考译文

1. 随着云计算的出现,为了简化网络管理和通过网络编程化带来创新,引入了许多新的网络概念。软件定义网络SDN实例的出现就是云模式中的概念之一,其目的是消除网络结构的维护过程和保证更容易的网络管理。

2. 在SDN方式中,SDN提供实时性能并对高可用性要求进行响应。然而,这种

SDN 新应用要面临许多技术障碍，一些是 SDN 本身固有的，而其他障碍源于现有继承的技术。

3. 一方面 SDN 可以结合系统虚拟化和云计算的优势，另一方面为了方便网络管理和维护以及提高网络控制和反应性可以实现智能化中心。

4. 在 SDN 中，控制操作集中在一个控制器中，该控制器指示网络策略。许多控制器平台都是开源的，比如 Floodlight、OpenDaylight 和 Beacon。网络的管理可以在不同的层（即应用程序、控制和数据平面）。

5. 称作控制平面的第二层是网络拓扑结构的抽象结果。其中的控制器是主要部件，用来建立流表和数据处理策略以及对网络复杂性进行抽象，还可以通过南向 API 收集网络信息。

6. 称作数据平面的最下一层提供联网设备如实体/虚拟交换机、路由器和接入点，并且负责数据转发、分片和重组等所有数据活动。软件定义网络近年来逐渐为人所知。

7. 此外，对各种形式的云服务的需求也在急剧增加。尽管这些服务集中在数据中心，但它们对服务提供者构成了重要的挑战。

8. 由于 SDN 是一种新的网络方法，使用这种体系结构对传统网络问题的几种解决方案进行了重新探讨，许多问题仍然具有挑战性。我们试图简化和解释一些 SDN 问题，并提供了一个总体的观点。

UNIT 17
Introduction to Optical Fiber Communication

一、词组翻译参考答案

1. optical fiber communications
2. light source
3. wavelength
4. laser
5. dispersion
6. transmission medium
7. multi-mode fiber
8. long-haul trunks
9. single-mode fiber
10. bandwidth
11. wideband subscriber
12. fiber-optics
13. commercial technology

14. threshold current

15. photodetector

16. wavelength multiplexing

17. fiber-optic networks

18. video bandwidth

二、词组翻译参考答案

1. 长途传输

2. 中继距离

3. 商用技术

4. 光纤通信

5. 已装光纤的总长度

6. 长途通信系统

7. 低衰减的石英纤维

8. 衰减逼近瑞利极限的光纤

9. 室温下的门限电流

10. 较长波长区

11. 用户接入工程

12. 部件性能和可靠性的改进

13. 已安装的光纤系统的数据速率

14. 每秒吉比特

15. 波分复用

16. 宽带用户环路系统

17. 多纤连接器

18. 设计寿命

19. 光源

20. 单模光纤

21. 分布反馈式激光器

22. 信息容量

23. 交换体系

24. 宽带业务

三、选择题答案

1. D　2. C　3. A　4. B　5. C　6. D

四、选择题答案

1. A　2. A　3. A　4. B　5. C　6. D

五、参考译文

1. 虽然现在人们对纤维光学的兴趣主要在于通信,但早期发展纤维光学技术的目的并非如此。20世纪50年代初,研究人员制造出第一根具有包层的玻璃纤维时,并没有打算将它用于通信,他们想用它用于传送内窥镜的成像光束。1966年Kao和Hockham发表了那篇著名的论文，建议将低损耗光纤用于通信，此时纤维光学已发展成一项成熟的技术了。

2. 1970 年 10 月，第一根低损耗石英光纤问世。人们有时将这个日期作为光纤通信时代的开始。尽管这一成果在当时的研究领域确实引起了极大的关注，但若因此就认为一项产业会发展起来尚为时太早。

3. 20 dB/km 的损耗对于长途通信系统仍然太大；光纤易断裂，必须找出保护办法；没有合适的光源；研究人员尚不知道光缆的终端和接头是否实际可行。最后，对于这些部件是否能被足够经济地生产出来，以使之在市场中发挥重要作用，他们更是心存疑虑。

4. 20 世纪 70 年代的中后期，由于发展重点由研究领域转向工程，因而加速了产品推向市场的速度。在实验室研制的光纤衰减值逼近了瑞利极限。通过改进光纤外涂层方法和成缆技术，克服了微弯损耗。生产出了工作于 1.3 μm 波长的光源和改进的光电检测器，从而可以利用这个“较长波长区”的低衰耗和低色散。

5. 最近敷设的光纤系统的数据速率已移至每秒吉比特范围，这种系统采用光谱纯的分布反馈激光器，将光纤色散效应减至最小。在 1.55 μm 波长上设计的低色散光纤，相应地具有低损耗特性，目前广泛地用于长途通信。为进一步增加光纤的信息容量，波分复用正在广泛应用。

6. 人们对于光纤在其他领域的潜力刚刚开始认识。用于计算机系统和办公室的光纤网络逐步变得更加重要。在电话系统中，光纤在电话交换系统低层的应用仍在迅猛发展。入户光缆已有了示范工程。许多观察家相信，全国电话系统将通过使用光纤而逐步升级到传输视频宽带信号。

7. 在基本光纤系统中包括 3 个部件：光源、光检测器和光传输线路。光源产生光能，它作为信息的载体，就像无线电波提供在某个无线波长上的电磁能量作为信息载体一样。光检测器检测光的能量并将其转换成电的形式。光纤传输线路与铜线等效，用作光的传导体。